BIGGER
BOLDER
BAKING
Every Day

BIGGER BOLDER BAKING
Every Day

Easy Recipes to Bake Through a Busy Week

GEMMA STAFFORD

HARVEST

An Imprint of WILLIAM MORROW

HarperCollins books may be purchased for educational, business, or sales promotional use. For information, please email the Special Markets Department at SPsales@harpercollins.com.

FIRST EDITION

Designed by Tai Blanche
Photography © 2022 by Carla Choy
Culinary Assistant Ami Shukla
Prop Styling by Kate Martindale and Olivia Crouppen

Library of Congress Cataloging-in-Publication Data has been applied for.

ISBN 978-0-358-46120-3

22 23 24 25 26   RTL   10 9 8 7 6 5 4 3 2 1

Kevin, this is for you.
"Whatever it takes!"

Love,
Gemma

TABLE OF CONTENTS

x   Introduction
xiii   Pro Chef Baking Tips
xiv   Useful Baking Information

CHAPTER 1

BREAKFAST IN MINUTES

3   Traditional Irish Scones
4   Cinnamon Apple Scones
7   Raspberry and Yogurt Scones
8   Maple Pecan Scones
10   Three-Seed Whole Wheat Scones
13   Strawberry Cream Cheese Scones
14   Bakery-Style Lemon Blueberry Muffins
17   Triple Chocolate Muffins
18   Sour Cream Coffee Cake Muffins
20   Banana Bread Muffins
25   Pumpkin Chocolate Chip Muffins
27   Apple Oatmeal Muffins
28   All-American Flaky Biscuits
30   Buttermilk Drop Biscuits
33   Whole Wheat and Fruit Breakfast Bread
34   Carrot and Marmalade Bread

CHAPTER 2

AFTERNOON TEA

39   Mia's Buttery Shortbread Cookies
41   Authentic Linzer Cookies
42   Homemade Chocolate-Oat Tea Cookies
44   Classic Chocolate Éclairs
47   Cappuccino Swiss Roll

49   Chocolate Swirl Meringues
50   Moist Fruit-and-Nut Cake
52   Mum's Irish Apple Cake
55   Citrus Olive Oil Pound Cake
56   All-American Coffee Crumb Cake
59   French Yogurt Pot Cake
60   Ami's Double Chocolate Almond Cookies
62   Butterscotch Banana Cake
65   Polenta Cake with Mascarpone and Strawberry Compote

CHAPTER 3

WEEKNIGHT FAMILY FAVORITES

68   10-Minute Summer Berry Tiramisu
70   Old-Fashioned Banana Pudding
73   Steamed Marmalade Pudding
75   Flaky Peach Shortcakes
76   Whole Lemon Tart
79   Chocolate Lover's Cheesecake with Strawberry Compote
81   Baked Custard with Rhubarb Compote
82   Smashed Raspberry Pavlova
85   Squidgy Chocolate Cake
87   Traditional Irish Bread-and-Butter Pudding
88   Classic Blueberry Pie
90   The Ultimate White Chocolate Pecan Skillet Cookie
92   Cool and Creamy Lime Custard Pie
95   Cinnamon Semifreddo with Honey Toffee Swirl

97 Strawberry Dump Cake

98 Pecan Pie Cobbler

100 Gooey Jam Tart

CHAPTER 4

DINNER PARTY DESSERTS

104 White Chocolate and Passion Fruit Cheesecake

106 Yogurt Coeur à la Crème with Macerated Cherries

108 Decadent Cocoa Panna Cotta

111 Elegant Tiramisu Crepe Cake

112 Chèvre and Honey Tart with Blueberry Compote

115 Boozy Chocolate and Prune Cake

116 Sparkling Rosé and Raspberry Granita

118 Dulce de Leche Lava Cake

121 Silky Coffee Pudding

122 Coconut Crème Brûlée

124 Grown-Ups' Rum Raisin Semifreddo

126 Death by Chocolate Cake

131 Layered Pavlova with Summer Fruit and Rose

132 Unbelievable Pear and Dark Chocolate Crisp

CHAPTER 5

WEEKEND BRUNCH TREATS

136 No-Yeast Cinnamon Rolls

139 Overnight Belgian Waffles

141 Banana Cinnamon Waffles

142 5-Ingredient Dutch Baby Pancake

145 Carrot Cake Pancakes with Cream Cheese Frosting

147 Kevin's No-Fuss Banana Oat Pancakes

149 Classic Popovers

150 Make-Ahead French Toast Casserole

153 Old-Fashioned Johnnycakes

155 Lemon-Blueberry Ricotta Hotcakes

157 Georgie's Aussie Pikelets

158 Raspberry and Cream Cheese Crepe Casserole

CHAPTER 6

LEISURELY WEEKEND RECIPES

162 No-Knead Cinnamon Raisin Bread

165 Sweet Pumpernickel Bread

166 Braided Chocolate Babka

170 Caramel Pecan Monkey Bread

175 My Go-To Brioche Loaf

177 Breakfast Brioche Buns

178 Easy-Peasy Homemade Croissants

183 Sticky Maple Walnut Morning Buns

185 Almond Twist

186 Buttery Fruit Danishes

188 Hawaiian Sweet Rolls

CHAPTER 7

SHORT AND SWEET ANY DAY

192 Fruit Fools

192 Strawberry and Mint Fool

192 Blueberry and Rose Water Fool

194 Boozy Raspberry Fool

194 Summer Blackberry Fool

195 Fruit Crisps

195 Triple Berry Crisp

195 Traditional Apple Crisp

197 Blueberry Peach Crisp

197 Rhubarb and Strawberry Crisp

198 Clafoutis

198 Clafouti Batter

199 Cherry Amaretto Clafoutis

199 Peach and Raspberry Clafoutis

201 Blackberry Clafoutis

201 Fresh Fig and Cardamom Clafoutis

202 Pan-Roasted Fruit

202 Roasted Pineapple with Rum Coconut Sauce

204 Roasted Strawberries with Rose and Cardamom Yogurt

205 Honey Glazed Figs with Whipped Mascarpone

205 Caramelized Bananas with Rosemary

206 Crostatas

206 Chocolate, Hazelnut, and Mascarpone Crostata

207 Caramel Apple Crostata

209 Berry and Cream Cheese Crostata

210 Plum, Crème Fraîche, and Vanilla Bean Crostata

211 Upside-Down Cakes

211 Upside-Down Cake Batter

212 Maple and Pear Upside-Down Cake

214 Orange and Honey Toffee Upside-Down Cake

215 Banana and Rum Upside-Down Cake

216 Fall Apple Cinnamon Upside-Down Cake

CHAPTER 8

MASTER RECIPES

220 Basic Meringue

222 Pie Crust

225 Crepes

226 Crisp Topping

228 Danish Dough

230 Honey Toffee Sauce

232 Whipped Cream

234 Fruit Compotes

234 Strawberry Compote

234 Blueberry and Lemon Compote

236 Rhubarb Compote

236 Raspberry Vanilla Compote

237 Chocolate-Butter Glaze

238 Baking FAQs

242 Acknowledgments

245 Index

INTRODUCTION

Hi, Bold Bakers!

Thank you so much for picking up my second cookbook! I'm so excited to share my new recipes; some of them are old and beloved, some new and chock-full of flavor, and all come with tips, tricks, and twists along the way!

Writing a book is telling a story, and sometimes, when telling a story, you get lost somewhere between the beginning and the end. When I first started writing this book, I had some ideas about what I wanted to include: ideas for chapter titles, ideas for a few essential recipes I *had* to include and others I wanted to perfect before sharing. Beyond that, I couldn't see clearly what the book would look like when it all came together.

From my experience writing my first book, *Bigger Bolder Baking: A Fearless Approach to Baking Anytime, Anywhere*, I knew that through the writing process and plenty of recipe testing, this book would find its way into creation. Lo and behold, it happened, and I could not be prouder of the results. Somewhere during the days spent bent down in front of the oven, late nights of deadline anxiety, and trying to get my neighbors, friends, and family to keep accepting many, many test desserts, this book came to life!

As the book started to unfold in front of me, I realized that what you start out planning and what you ultimately end up with can be two very different things, which is really exciting. At first, I had the intention of writing a book that was about baking throughout the week.

What I ended up with is a comprehensive collection of desserts and sweets for EVERY DAY! These recipes were all curated so that no matter the occasion or time constraints, there is the perfect recipe in this book for you—from Breakfast in Minutes to sophisticated Dinner Party Desserts and everything in between!

As a professional baker with twenty years of experience, I'm thrilled to be able to provide you with an extensive repertoire of recipes made to be baked any day of the week. I hope some of my favorites become something you and your family can share for years to come.

Why This Book

This collection of recipes includes everything from bread doughs to fancy desserts. From pancakes (see page 142) to crème brûlée (see page 122) and everything in between. This is something my Bold Bakers have been asking for time and time again over the years. Consider it a go-to one-stop shop for recipes! You can find whatever you need in here, be it a basic Danish Dough (page 228), Traditional Irish Scones (page 3), or an elegant Layered Pavlova (page 131)!

How to Use This Book

Using this book is as simple as some of the recipes you'll find in it. This book is broken into chapters by time and occasion. Just choose what type of day you're having, and get ready to bake up anything from a quick weeknight dessert to a more time-intensive bread!

PRO CHEF BAKING TIPS

Here are some tips I have gathered over the years as a professional chef that I still find useful in my kitchen every day.

- Have an abundance of excess egg whites? Pour them into zip-top bags, label with the number of egg whites, and freeze for up to 4 months. Defrost at room temperature and use as you normally would.

- A large egg in its shell weighs 60 grams. Once cracked that gives you 30 grams = white and 20 grams = yolk. Remember this if you need to measure eggs by weight.

- When making meringues and pavlovas, always have your egg whites at room temperature because they will whip so much better than when they're cold from the fridge.

- Pop cold eggs in a bowl of warm water to bring them to room temp.

- Bits of shell in your eggs? Use half an eggshell to scoop it out. The pieces will be drawn to each other like magnets.

- If you're making a custard and it splits, don't fear! You can bring it back together with an immersion blender in seconds and it will be good as new.

- When making chocolate chip cookies, buy a solid bar of chocolate and chop it into pieces instead of using chocolate chips. The pieces melt better and the chocolate is usually of better quality.

- Always store whole wheat flour in the fridge. This keeps it fresh and prevents it from going rancid.

- Save your butter wrappers and use them to grease your baking pans. It's a great way to reuse them and means you don't need to dirty a pastry brush.

- Put vegetable or olive oil in a spray bottle to use for greasing bowls when making bread.

- When making bread, put yeast on one side and salt on the other when you add them to the flour in the bowl. Don't place them directly on top of each other or the salt can inhibit the yeast from activating.

- For speedy proofing of dough, preheat the oven for 5 minutes, then turn it off and pop in your dough. The warmth will help it proof faster.

- Use a shower cap to cover bowls of dough before proofing. They are cheap and reusable and cover bowls tightly to keep out air. For extra warmth, place a clean kitchen towel over the shower cap.

- Store dry yeast in a tub in the fridge to keep it fresh for much, much longer.

- When shaping bread rolls, DON'T flour your work surface. Your dough will be easier to roll if it can get traction on the surface.

- What does "blood-temperature water" mean when making bread? Put your fingers in a jug of warm water. If you can't feel the water around your finger, it means it's the same temperature as your blood and the perfect temperature for bread making.

- Went crazy making bread? Slice the excess or leftover loaf, but keep all the slices together. Place them in a bag and freeze so you have easy access to bread for toast.

USEFUL BAKING INFORMATION

Baking Conversion Chart

ALL-PURPOSE FLOUR

¼ cup	1¼ ounces	35 grams
⅓ cup	1½ ounces	43 grams
½ cup	2½ ounces	71 grams
⅔ cup	3⅓ ounces	94 grams
¾ cup	3¾ ounces	105 grams
1 cup	5 ounces	142 grams

CONFECTIONERS' SUGAR

¼ cup	1 ounce	28 grams
⅓ cup	1⅓ ounces	37 grams
½ cup	2 ounces	57 grams
⅔ cup	2⅔ ounces	76 grams
¾ cup	3 ounces	85 grams
1 cup	4 ounces	115 grams

GRANULATED SUGAR

¼ cup	2 ounces	57 grams
⅓ cup	2½ ounces	71 grams
½ cup	4 ounces	115 grams
⅔ cup	5 ounces	142 grams
¾ cup	6 ounces	170 grams
1 cup	8 ounces	225 grams

COCOA POWDER

¼ cup	1 ounce	28 grams
⅓ cup	1⅓ ounces	37 grams
½ cup	2 ounces	57 grams
⅔ cup	2⅔ ounces	76 grams
¾ cup	3 ounces	85 grams
1 cup	4 ounces	115 grams

BROWN SUGAR (NOT PACKED)

¼ cup	1½ ounces	43 grams
⅓ cup	2 ounces	57 grams
½ cup	3 ounces	85 grams
⅔ cup	4 ounces	115 grams
¾ cup	4½ ounces	128 grams
1 cup	6 ounces	170 grams

BUTTER

4 tablespoons (½ stick)	2 ounces	57 grams
⅓ cup	2½ ounces	71 grams
½ cup (1 stick)	4 ounces	115 grams
⅔ cup	5 ounces	142 grams
¾ cup (1½ sticks)	6 ounces	170 grams
1 cup (2 sticks)	8 ounces	225 grams

CHOCOLATE
(CHIPS AND CHOPPED BAR)

¼ cup	1½ ounces	43 grams
⅓ cup	2 ounces	57 grams
½ cup	3 ounces	85 grams
⅔ cup	4 ounces	115 grams
¾ cup	4½ ounces	128 grams
1 cup	6 ounces	170 grams

FRUIT (FRESH OR FROZEN)

¼ cup	1¼ ounces	35 grams
⅓ cup	1½ ounces	43 grams
½ cup	2½ ounces	71 grams
⅔ cup	3⅓ ounces	94 grams
¾ cup	3¾ ounces	106 grams
1 cup	5 ounces	142 grams

NUTS

¼ cup	1¼ ounces	35 grams
⅓ cup	1½ ounces	43 grams
½ cup	2½ ounces	71 grams
⅔ cup	3⅓ ounces	94 grams
¾ cup	3¾ ounces	106 grams
1 cup	5 ounces	142 grams

DRIED FRUIT

¼ cup	1¼ ounces	35 grams
⅓ cup	1½ ounces	43 grams
½ cup	2½ ounces	71 grams
⅔ cup	3⅓ ounces	94 grams
¾ cup	3¾ ounces	106 grams
1 cup	5 ounces	142 grams

HONEY/MAPLE SYRUP/AGAVE NECTAR

¼ cup	2½ ounces	71 grams
⅓ cup	4 ounces	115 grams
½ cup	5 ounces	142 grams
⅔ cup	7½ ounces	213 grams
¾ cup	8 ounces	225 grams
1 cup	10 ounces	283 grams

LIQUID

¼ cup	2 fluid ounces	60 milliliters
⅓ cup	2½ fluid ounces	75 milliliters
½ cup	4 fluid ounces	120 milliliters
⅔ cup	5 fluid ounces	150 milliliters
¾ cup	6 fluid ounces	180 milliliters
1 cup	8 fluid ounces	240 milliliters

Oven Temperature Chart

FAHRENHEIT AND CELSIUS

°F	°C	GAS MARK	
500°F	260°C	10	Extremely hot—pizza
475°F	240°C	9	Very hot—flatbread, pizza
450°F	230°C	8	Very hot—some bread
425°F	220°C	7	Hot—good for simple yeast bread
400°F	200°C	6	Hot—roasting; some enriched bread, some pastry
375°F	190°C	5	Moderately hot—some cookies, pastries, and bread
350°F	180°C	4	Moderate—good baking temperature; cakes, pastry
325°F	170°C	3	Warm—good for baked custards, e.g., crème caramel
300°F	150°C	2	Gentle—good for French macarons, for example
275°F	140°C	1	Cool—good for meringue, pavlova
250°F	130°C	½	Low and slow—drying; dehydrating fruits and vegetables
225°F	110°C	¼	Very low and slow—drying herbs, vanilla beans, citrus

Using a convection oven: The general rule if you're using a convection oven with the fan on is to set the oven temperature 20°F (10°C) lower than what's called for in recipes using a regular oven.

Cooking time: Don't go strictly by the cooking time indicated in a recipe, as ovens vary; instead, be sure to check on your food regularly to see how fast it is cooking.

Common Baking Pan Sizes

Round Cake Pans
6 x 2 inches (15 x 5 cm)
8 x 2 inches (20 x 5 cm)
9 x 2 inches (23 x 5 cm)

Square Pans
8 x 8 inches (20 x 20 cm)
9 x 9 inches (23 x 23 cm)

Rectangular Pans
11 x 7 inches (28 x 18 cm)
13 x 9 inches (33 x 23 cm)

Springform Pans
9 x 2.5 inches (23 x 6 cm)
10 x 2.5 inches (25 x 6 cm)

Bundt Pan (volume varies because of various designs)
10 x 3 inches (25 x 8 cm)

Jelly-Roll Pan
10 x 15 inches (27 x 39 cm)

Loaf Pans
8 x 4 inches (20 x 10 cm)
9 x 5 inches (23 x 13 cm)

Chapter 1

BREAKFAST in MINUTES

In Ireland, our days are fueled by scones and a good cup of tea first thing in the morning. It's a habit I have not grown out of, nor do I want to! There's no better way to start your day than with a warm, freshly baked treat.

These scones, muffins, and quick breads can all be made by hand, so there is no need for any heavy equipment. Grab a bowl and tie your apron around your waist, and before the kettle has had time to whistle, your breakfast will be coming out hot from the oven.

TRADITIONAL IRISH SCONES

Ireland is a nation of tea drinkers, and I wouldn't be able to visit home again if I didn't include a recipe in this book for traditional Irish scones! The humble scone is a huge part of day-to-day life in Ireland, and I've spent years looking high and low for, working on, and perfecting this Irish scone recipe.

Brew yourself a cup of tea while these are still warm, and don't be shy with the jam and cream. I think I'm making my people (and my mum!) proud with this one!

MAKES 10 SCONES

4⅔ cups (680 grams) self-rising flour

½ cup (115 grams) granulated sugar

1 level tablespoon baking powder

¾ cup (1½ sticks/170 grams) butter, frozen

1 cup (142 grams) raisins

2 large eggs, at room temperature

1¼ cups (300 milliliters) whole milk

1 egg, beaten, for egg wash

Butter, jam, or Whipped Cream (page 232), for serving

1 Preheat the oven to 425°F (220°C). Line two baking sheets with parchment paper.

2 In a large bowl, mix together the flour, sugar, and baking powder.

3 Using the large holes of a box grater, grate the frozen butter directly into the dry ingredients until it is all gone and quickly toss it in the flour to separate the pieces. The mixture should be somewhat crumbly. Stir in the raisins.

4 In a measuring cup, whisk together the eggs and milk until thoroughly combined. Pour the milk mixture into your flour mixture and stir until a soft dough is formed. (If the ingredients are not coming together into a dough or seem a little dry, add a little more liquid.)

5 Turn the dough out onto a floured surface and press it into a layer about 1½ inches thick. Use a 3-inch round cookie cutter to cut out as many scones as you can, placing them on the prepared baking sheets as you go.

6 Gather the scraps of dough into a ball, flatten, and cut more scones. Repeat until you have used all the dough. Brush the tops of the scones with the egg wash.

7 Bake for 22 to 26 minutes, until golden brown.

8 Serve warm, with butter, jam, or whipped cream. Store leftover scones in an airtight container at room temperature for up to 2 days.

CINNAMON APPLE SCONES

I wouldn't say I like change, especially when it comes to my food. I'm a human being, and I'm not ashamed to say that there was a time in my life when if you added anything but raisins to a scone, I would have walked out the door! But when I left Ireland, I realized I was missing out on a whole world of scones.

Look at these cinnamon apple scones! With the apple-pie-like filling and the sweet cinnamon glaze on top, how could you say no? Serve them as they are—no butter, jam, or whipped cream needed.

MAKES 8 SCONES

FOR THE APPLE FILLING

- 1 Granny Smith apple (about 5 ounces/142 grams), cored, peeled, and finely diced
- 1 tablespoon granulated sugar
- 1 teaspoon ground cinnamon

FOR THE DOUGH

- 2 cups (284 grams) all-purpose flour
- ¼ cup (57 grams) granulated sugar
- 1 teaspoon baking powder
- ¼ teaspoon baking soda
- ½ teaspoon salt
- ½ cup (1 stick/115 grams) cold butter, diced
- 1 cup (240 milliliters) cold heavy cream
- 1 egg, beaten, for egg wash

FOR THE GLAZE

- 1 cup (115 grams) confectioners' sugar
- 2 tablespoons whole milk
- 1 teaspoon ground cinnamon
- ½ teaspoon pure vanilla extract

1 Preheat the oven to 400°F (200°C). Line a baking sheet with parchment paper.

2 **Make the apple filling:** In a small bowl, combine the apples, granulated sugar, and cinnamon and set aside.

3 **Make the dough:** In a large bowl, whisk together the flour, granulated sugar, baking powder, baking soda, and salt.

4 Add the butter and, using your fingertips, rub it into the dry ingredients until the mixture resembles coarse bread crumbs.

5 Add the cream and stir until the dough comes together. (If your scones are not forming a dough and seem a little dry, then add a little more liquid.)

6 Turn the dough out onto a floured surface and roll it into a ¼-inch-thick rectangle. Spread the apple filling evenly over the surface.

7 Fold the dough in thirds like a letter and roll it out into a 9 x 4-inch rectangle, roughly 1 inch thick. Cut into 8 triangles and place on the prepared baking sheet. Brush with the egg wash.

8 Bake for about 20 minutes, until the tops are golden brown. Let cool.

9 **Make the glaze:** In a small bowl, whisk together the confectioners' sugar, milk, cinnamon, and vanilla. Drizzle the glaze over the scones and let it set before serving.

10 Serve with a hot cup of tea. Store leftovers in an airtight container at room temperature for up to 2 days.

RASPBERRY and YOGURT SCONES

When I create recipes, there always has to be a reason. So what's the reason for my yogurt scones? It's all about how the ingredients perfectly complement each other not just in taste but in the way they affect one another. Yogurt is naturally acidic, and that acid helps tenderize the flour. I dare you to find a softer scone!

As for the "why raspberry"? That's a no-brainer—raspberry jam is my favorite!

MAKES 8 SCONES

FOR THE RASPBERRY FILLING

- 1 cup (5 ounces/142 grams) fresh raspberries
- 3 tablespoons raspberry jam, warmed slightly

FOR THE DOUGH

- 2 cups (284 grams) all-purpose flour
- ⅓ cup (71 grams) granulated sugar
- 2 teaspoons baking powder
- ¼ teaspoon baking soda
- ½ teaspoon salt
- ½ cup (1 stick/115 grams) cold butter, diced
- ⅓ cup (71 grams) plain yogurt
- ⅓ cup (71 milliliters) heavy cream
- 1 large egg, at room temperature
- 1 egg, beaten, for egg wash
  Coarse sugar, for decorating

1 Preheat the oven to 400°F (200°C). Line a baking sheet with parchment paper.

2 **Make the raspberry filling:** In a small bowl, gently combine the raspberries and jam. Set aside.

3 **Make the dough:** In a large bowl, whisk together the flour, sugar, baking powder, baking soda, and salt.

4 Add the butter and, using your fingertips, rub it into the dry ingredients until the mixture resembles coarse bread crumbs.

5 In a small bowl, whisk together the yogurt, cream, and egg, then pour this into the flour mixture and stir until the dough comes together. (If the mixture is not forming a dough and seems a little dry, add a little more liquid.)

6 Turn dough out onto a floured surface and divide the dough in half and roll each half into a circle about 8 inches across and ¼ inch thick.

7 Place one circle of dough on the prepared baking sheet and spread it with the filling, leaving a ½-inch border. Place the second circle of dough on top of the first and press the edges to seal in the filling.

8 Using a knife, score the disc into 8 wedges, cutting about three-quarters of the way through to the bottom. Brush the tops with the egg wash and sprinkle with coarse sugar.

9 Bake for 25 to 30 minutes, until golden brown.

10 Serve as is, warm from the oven. Store leftovers in an airtight container at room temperature for up to 2 days.

MAPLE PECAN SCONES

A couple of fun facts for you: Pecans are the only major tree nut native to America, the US produces around 300 million tons of pecans every year, and *they are my favorite nut!*

Pecans are magic when it comes to baking. Raw pecans have a buttery flavor, but my chef's secret is to always toast them. They take on a lovely maple taste with caramel tones when toasted, flavors that work so well in many different recipes.

MAKES 8 SCONES

FOR THE DOUGH

2	cups (284 grams) all-purpose flour
1	level tablespoon baking powder
¼	teaspoon baking soda
¼	teaspoon salt
½	cup (1 stick/115 grams) cold butter, diced
1	cup (142 grams) pecans, toasted and chopped
½	cup (115 grams) sour cream
¼	cup (60 milliliters) buttermilk
¼	cup (71 milliliters) pure maple syrup
1	egg, beaten, for egg wash

FOR THE MAPLE GLAZE

¾	cup (85 grams) confectioners' sugar
⅓	cup (115 grams) pure maple syrup
1	tablespoon (14 grams) butter, melted

1 Preheat the oven to 400°F (200°C). Line a baking sheet with parchment paper.

2 **Make the dough:** In a large bowl, whisk together the flour, baking powder, baking soda, and salt.

3 Add the butter and, using your fingertips, rub it into the dry ingredients until the mixture resembles coarse bread crumbs. Toss in the pecans and mix to combine.

4 In a small bowl, combine the sour cream, buttermilk, and maple syrup, then pour this into the flour mixture and stir until the dough comes together. (If the mixture is not forming a dough and seems a little dry, add a little more liquid.)

5 Turn the dough out onto a floured surface and roll it out to 1 inch thick. Use a 3-inch round cookie cutter to cut out as many scones as you can, placing them on the prepared baking sheet as you go.

6 Gather the scraps of dough into a ball, roll out the dough again, and cut out more scones. Repeat until you have used all the dough.

7 Brush the tops of the scones with the beaten egg wash.

8 Bake for 25 to 30 minutes, until the tops are golden brown. Let cool.

9 **Make the glaze:** In a small bowl, whisk together the confectioners' sugar, maple syrup, and butter.

10 Drizzle the glaze over the scones and let it set before enjoying. Store leftovers in an airtight container at room temperature for up to 2 days.

THREE-SEED WHOLE WHEAT SCONES

You can have the best of both worlds with this recipe! Whole wheat flour is nutritious and adds a delicious nuttiness that is really unique. Try topping your scones with a drizzle of honey—it sweetens them and enhances the flavor!

MAKES 9 SCONES

1 cup (142 grams) whole wheat flour

1 cup (142 grams) all-purpose flour

2 tablespoons sesame seeds

2 tablespoons hulled pumpkin seeds

2 tablespoons hulled sunflower seeds

2 teaspoons baking powder

½ teaspoon baking soda

½ teaspoon salt

½ cup (1 stick/115 grams) cold butter, diced

½ cup (120 milliliters) heavy cream

3 tablespoons honey

1 large egg, at room temperature

1 egg, beaten, for egg wash
Butter, for serving

1 Preheat the oven to 375°F (190°C). Line a baking sheet with parchment paper.

2 In a large bowl, combine the whole wheat flour, all-purpose flour, sesame seeds, pumpkin seeds, sunflower seeds, baking powder, baking soda, and salt.

3 Add the butter and, using your fingertips, rub it into the dry ingredients until the mixture resembles coarse bread crumbs.

4 In a small measuring cup, whisk together the cream, honey, and egg, then pour this into the flour mixture and stir until the dough comes together. (If the mixture is not forming a dough and seems a little dry, add a little more liquid.)

5 Turn the dough out onto a floured surface and roll it out into a 9-inch square, roughly ¾ inch thick. Using a knife, cut the dough into 3-inch squares (you should end up with 9 scones) and place them on the prepared baking sheet. Brush the tops with the egg wash.

6 Bake for 20 to 25 minutes, until golden brown. Let cool for 10 minutes before serving.

7 Serve the scones warm, with butter. Store leftovers in an airtight container at room temperature for up to 2 days.

STRAWBERRY CREAM CHEESE SCONES

Once I started experimenting with scone recipes, I couldn't stop! Did you know there are no rules you have to follow? Because there are no rules when it comes to scones, you can do whatever you want!

Take this strawberry cream cheese scone recipe, for example. It is as tender as a buttery biscuit, has a perfect slight tanginess from the cream cheese, and is filled with fresh, juicy strawberries. It's the perfect combination of flavor and texture—these scones are the complete package!

MAKES 8 SCONES

FOR THE STRAWBERRY FILLING

- 1 cup (5 ounces/ 142 grams) fresh strawberries, chopped
- 1½ teaspoons all-purpose flour

FOR THE DOUGH

- 2¼ cups (319 grams) all-purpose flour
- ¼ cup (57 grams) granulated sugar
- 1 level tablespoon baking powder
- ½ teaspoon salt
- ½ cup (115 grams) cream cheese, diced
- ½ cup (1 stick/115 grams) cold butter, diced
- ¼ cup (60 milliliters) heavy cream
- 1 large egg, at room temperature
- 1 egg, beaten, for egg wash

1 Preheat the oven to 400°F (200°C). Line a baking sheet with parchment paper.

2 **Make the strawberry filling:** In a small bowl, toss the strawberries with the flour.

3 **Make the dough:** In a large bowl, whisk together the flour, sugar, baking powder, and salt.

4 Add the cream cheese and butter and, using your fingertips, rub them into the dry ingredients until the mixture resembles coarse bread crumbs.

5 In a small bowl, beat the heavy cream and egg together, then pour this into the flour mixture and stir until the dough comes together. (If the mixture is not forming a dough and seems a little dry, add a little more liquid.)

6 Transfer the dough to a floured surface and roll it out into a 9 x 12-inch rectangle, roughly ¼ inch thick. Spread the strawberry filling evenly over the dough.

7 Fold the dough into thirds, like a letter, and roll it out into a 9 x 4½-inch rectangle, about 1 inch thick. Cut the rectangle into 8 squares and place them on the prepared baking sheet. Brush the tops with the egg wash.

8 Bake for 20 to 25 minutes, until golden brown.

9 Let cool slightly before enjoying. Store leftovers in an airtight container at room temperature for up to 2 days.

BAKERY-STYLE LEMON BLUEBERRY MUFFINS

When I started to work on my lemon blueberry muffins, I wanted to make sure that they not only looked like real-deal bakery-style muffins but had the flavor and texture to match.

I'm happy to say these muffins are *bursting* with flavor—and the best part is, they only get better! The following day, the flavors develop, and they get that perfect sticky top we know and love. Bake these muffins the night before, and you're ready to go in the morning!

MAKES 8 MUFFINS

⅔ cup (142 grams) granulated sugar

½ cup (120 milliliters) vegetable oil

⅓ cup (71 milliliters) whole milk

¼ cup (57 grams) plain yogurt

1 teaspoon pure vanilla extract

1 tablespoon grated lemon zest

1 large egg, at room temperature

1¼ cups (177 grams) all-purpose flour

1 teaspoon baking powder

¼ teaspoon salt

1 cup (5 ounces/ 142 grams) blueberries, fresh or frozen

1 Preheat the oven to 375°F (190°C). Line 8 wells of a 12-cup muffin tin with paper liners.

2 In a medium bowl, whisk together the sugar, oil, milk, yogurt, vanilla, lemon zest, and egg until smooth.

3 In a large bowl, mix together the flour, baking powder, and salt.

4 Add the wet ingredients to the dry ingredients and stir until just combined. A few lumps are okay. Gently fold in the blueberries.

5 Divide the batter evenly among the prepared muffin cups, filling the liners almost to the top.

6 Bake for 35 to 40 minutes, until the muffins are golden brown on top.

7 Enjoy the muffins warm, just as they are. Store leftovers in an airtight container at room temperature for up to 3 days.

TRIPLE CHOCOLATE MUFFINS

This muffin is for everyone who craves chocolate for breakfast, featuring semisweet chocolate, bittersweet chocolate, and dark cocoa powder. Chocolate is truly the star of the show here—I couldn't fit more into this muffin, even if I tried!

MAKES 10 MUFFINS

- ½ cup (71 grams) all-purpose flour
- ⅓ cup (37 grams) unsweetened cocoa powder
- ½ teaspoon baking soda
- ½ teaspoon salt
- ⅔ cup (4 ounces/115 grams) chopped bittersweet chocolate, melted
- ½ cup (120 milliliters) vegetable oil
- ⅔ cup (115 grams) dark brown sugar
- ⅓ cup (71 grams) plain yogurt
- 2 large eggs, at room temperature
- 2 teaspoons pure vanilla extract
- ¾ cup (4½ ounces/ 128 grams) semisweet chocolate chips

1 Preheat the oven to 375°F (190°C). Line 10 wells of a 12-cup muffin tin with paper liners.

2 In a large bowl, whisk together the flour, cocoa powder, baking soda, and salt.

3 In a medium bowl, stir together the melted chocolate, oil, brown sugar, yogurt, eggs, and vanilla.

4 Add the wet ingredients to the dry ingredients and stir until just combined. Stir in roughly three-quarters of the chocolate chips, reserving the remainder for topping.

5 Divide the batter evenly among the prepared muffin cups, filling the liners almost to the top. Top the muffins with the remaining chocolate chips.

6 Bake for 17 to 18 minutes, until well risen and firm to the touch. Be careful not to overbake these guys so they stay moist in the middle.

7 Enjoy warm and melty, fresh from the oven. Store leftovers in an airtight container at room temperature for up to 3 days.

SOUR CREAM COFFEE CAKE MUFFINS

Don't be fooled by their size—these little sour cream coffee cake muffins pack a flavorful punch! They have all the characteristics that make classic coffee cake irresistible, including a sweet swirl of cinnamon and toasted walnuts inside. Of course, I didn't forget the iconic crumble on top for added texture; after all, the crumble topping is arguably the best part of a coffee cake. The sour cream also lends a hand in this recipe, as its acidity works its magic on the flour to make these muffins incredibly moist.

MAKES 12 MUFFINS

FOR THE STREUSEL TOPPING

- ½ cup (71 grams) all-purpose flour
- ½ cup (85 grams) dark brown sugar
- ½ teaspoon ground cinnamon
- 4 tablespoons (½ stick/57 grams) butter, softened

FOR THE BATTER

- 2 cups (284 grams) all-purpose flour
- 2 teaspoons baking powder
- ½ teaspoon baking soda
- 1 teaspoon ground cinnamon
- ¼ teaspoon salt
- 1 cup (225 grams) sour cream
- 1 cup (170 grams) dark brown sugar
- ½ cup (1 stick/115 grams) butter, softened
- 2 large eggs, at room temperature
- 2 teaspoons pure vanilla extract

FOR THE WALNUT CINNAMON SWIRL

- 1 cup (142 grams) walnuts, toasted and chopped
- ⅓ cup (57 grams) dark brown sugar
- 2 tablespoons (28 grams) butter, softened
- 1 teaspoon ground cinnamon

1 Preheat the oven to 375°F (190°C). Line a 12-cup muffin tin with paper liners.

2 **Make the streusel topping:** In a small bowl, mix the flour, brown sugar, and cinnamon. Stir in the butter until combined. Set aside.

3 **Make the batter:** In a large bowl, whisk together the flour, baking powder, baking soda, cinnamon, and salt.

4 In another large bowl, whisk together the sour cream, brown sugar, butter, eggs, and vanilla.

5 Add the sour cream mixture to the flour mixture and stir until just combined.

6 **Make the walnut cinnamon swirl:** In a small bowl, mix together the walnuts, brown sugar, butter, and cinnamon.

7 Swirl the walnut mixture into the muffin batter, then divide the batter evenly among the 12 prepared muffin cups, filling the liners almost to the top. Sprinkle each muffin evenly with the streusel topping.

8 Bake for 18 to 20 minutes, until a toothpick inserted into the center comes out clean. Let cool in the muffin tin for 10 minutes before serving.

9 Enjoy while fresh and warm. Store leftovers in an airtight container at room temperature for up to 3 days.

BANANA BREAD MUFFINS

This banana bread muffin recipe is chock-a-block with bananas! There's no better way to use up any bananas you might have let get too brown on your counter.

May I offer a bit of advice? You don't want to skimp on the salt and vanilla extract in this one. Those two ingredients really complement the flavor of the bananas and will have you coming back for seconds (or thirds!).

MAKES 12 MUFFINS

1¾ cups (247 grams) all-purpose flour

1 teaspoon baking powder

1 teaspoon baking soda

½ teaspoon ground cinnamon

¼ teaspoon salt

4 ripe medium bananas (roughly 15 ounces/ 426 grams), peeled and mashed

1 cup (225 grams) granulated sugar

¾ cup (170 millimeters) vegetable oil

½ cup (115 grams) sour cream

2 large eggs, at room temperature

1 tablespoon pure vanilla extract

1 Preheat the oven to 350°F (180°C). Line a 12-cup muffin tin with paper liners.

2 In a large bowl, whisk together the flour, baking powder, baking soda, cinnamon, and salt.

3 In another large bowl, stir together the mashed bananas, sugar, oil, sour cream, eggs, and vanilla until well mixed.

4 Add the wet ingredients to the dry ingredients and stir until just combined.

5 Divide the batter evenly among the prepared muffin cups, filling the liners almost to the top.

6 Bake for 20 to 25 minutes, until a toothpick inserted into the center comes out clean. Let cool in the muffin tin for 10 minutes before serving.

7 Enjoy warm from the oven. Store leftovers in an airtight container at room temperature for up to 2 days.

PUMPKIN CHOCOLATE CHIP MUFFINS

Everyone gets excited when the air starts to get a little crisper and leaves begin to change their colors. Those are telltale signs that pumpkin season is back! But in my opinion, these gorgeous gourds need more time to shine than just a few months of the year.

Who says we can't eat these delicious pumpkin chocolate chip muffins all year long? Bust out this recipe any time you can't wait for autumn and the holidays to come!

MAKES 12 MUFFINS

1½ cups (213 grams) all-purpose flour

1 teaspoon ground cinnamon

1 teaspoon baking powder

½ teaspoon baking soda

½ teaspoon salt

1¼ cups (282 grams) granulated sugar

1 cup (8 ounces/ 225 grams) canned pure pumpkin puree

⅓ cup (71 millimeters) vegetable oil

2 large eggs, at room temperature

¾ cup (4½ ounces/ 128 grams) semisweet chocolate chips

1 Preheat the oven to 375°F (190°C). Line a 12-cup muffin tin with paper liners.

2 In a medium bowl, whisk together the flour, cinnamon, baking powder, baking soda, and salt.

3 In another medium bowl, whisk together the sugar, pumpkin, oil, and eggs until well combined.

4 Add the wet ingredients to the dry ingredients and stir until just combined.

5 Fold roughly three-quarters of the chocolate chips into the batter, reserving the remainder for topping.

6 Divide the batter evenly among the prepared muffin cups, filling the liners almost to the top. Sprinkle the tops with the remaining chocolate chips.

7 Bake for 25 to 30 minutes, until a toothpick inserted into the center comes out clean.

8 Enjoy warm from the oven. Store leftovers in an airtight container at room temperature for up to 3 days.

APPLE OATMEAL MUFFINS

My apple oatmeal muffins are filled with wonderful apple flavor and sweetened with honey, and the oatmeal gives them an incredible texture. They're one of my favorite go-to recipes; they make a great on-the-go breakfast, are perfect to include as a treat in a lunch box, or can even be simply eaten alfresco while you play with your toy cars. Baby George says pants are also optional while you're enjoying these muffins. (Can't argue with that logic!)

MAKES 12 MUFFINS

1½	cups (212 grams) all-purpose flour
½	cup (43 grams) rolled oats
2	teaspoons ground cinnamon
1	teaspoon baking powder
½	teaspoon baking soda
½	teaspoon salt
2	medium Granny Smith apples (about 10 ounces/284 grams), peeled, cored, and grated
⅔	cup (225 grams) honey
⅓	cup (71 milliliters) vegetable oil
2	large eggs, at room temperature
1½	teaspoons pure vanilla extract

1 Preheat the oven to 375°F (190°C). Line a 12-cup muffin tin with paper liners.

2 In a large bowl, whisk together the flour, oats, cinnamon, baking powder, baking soda, and salt.

3 In a separate large bowl, whisk together the apples, honey, oil, eggs, and vanilla.

4 Add the wet ingredients to the dry ingredients and stir until just combined.

5 Divide the batter evenly among the prepared muffin cups, filling the liners almost to the top.

6 Bake for 18 to 20 minutes, until a toothpick inserted into the center comes out clean.

7 Let cool slightly in the muffin tin before enjoying. Store leftovers in an airtight container at room temperature for up to 2 days.

ALL-AMERICAN FLAKY BISCUITS

What do I want from an authentic American biscuit? Layers, baby! Buttery, flaky layers or bust! I spent a lot of time testing recipe after recipe until finally nailing down what I think is the best biscuit recipe ever. I'll let you be the judge—but I won't be surprised if these biscuits become a common side dish at your dinner table.

MAKES 9 BISCUITS

- 4 cups (568 grams) all-purpose flour
- 2 level tablespoons baking powder
- 1 teaspoon salt
- 1 cup (2 sticks/225 grams) butter, frozen
- 1½ cups (340 milliliters) buttermilk
- 1 egg, beaten, for egg wash
- Butter and honey, for serving

1 Preheat the oven to 375°F (190°C). Line a baking sheet with parchment paper.

2 In a large bowl, whisk together the flour, baking powder, and salt.

3 Using the large holes of a box grater, grate the frozen butter directly into the dry ingredients and toss to combine.

4 Add the buttermilk and mix until the dough comes together. If it seems a little dry, add a splash more liquid.

5 Turn the dough out onto a lightly floured surface and press it into a rectangle about 1 inch thick.

6 Fold the dough in thirds, like a letter, then roll it out to about 1 inch thick.

7 Use a 3-inch round cookie cutter to cut as many biscuits as you can. Gather the scraps of dough into a ball, flatten, and cut more scones. Repeat until you have used all the dough. Brush the tops with the egg wash and place on the prepared baking sheet.

8 Bake for about 25 minutes, until golden brown.

9 Enjoy warm from the oven, with butter and honey. Store leftovers in an airtight container at room temperature for up to 3 days.

BUTTERMILK DROP BISCUITS

I've trained under acclaimed chefs and cookbook authors; pleased even the most unpleasant priest in my first professional pastry chef position at a priory in Dublin; and worked for years in fine dining, making Michelin-star-worthy desserts in San Francisco. But I still gravitate toward the simple recipes—maybe you're like me in that way, too! It doesn't get much simpler than this buttermilk drop biscuit. You get the buttery, tender crumb of a biscuit but without having to break out your rolling pin!

MAKES 10 BISCUITS

2 cups (284 grams) all-purpose flour

2 teaspoons baking powder

½ teaspoon baking soda

½ teaspoon salt

½ cup (1 stick/115 grams) cold butter, diced

1 cup (240 millimeters) buttermilk

1 egg, beaten, for egg wash

1 Preheat the oven to 475°F (240°C). Line a baking sheet with parchment paper.

2 In a large bowl, whisk together the flour, baking powder, baking soda, and salt. Add the butter and, using your fingertips, rub it into the dry ingredients until the mixture resembles coarse bread crumbs. Pour in the buttermilk and stir until just combined.

3 Using a ¼-cup (57-gram) measuring cup, drop 10 mounds of batter onto the prepared baking sheet, spacing them evenly. Brush the tops of the biscuits with the egg wash.

4 Bake for 12 to 16 minutes, until the tops are golden.

5 Let cool for 5 minutes before serving. Store leftovers in an airtight container at room temperature for up to 3 days.

WHOLE WHEAT *and* FRUIT BREAKFAST BREAD

In Ireland, my mum used to treat us occasionally on Friday mornings by bringing home "breakfast bread." It was dark, a little sweet, and full of dried fruit. It was perfect as it was or toasted and topped with a little salty butter to bring out even more sweetness. It's difficult to create a recipe based on something you loved so much (nostalgia will do that to you!), but this breakfast bread made with whole wheat flour and dried fruit takes me right back home—no flight required.

MAKES ONE 9 X 5-INCH LOAF

- 1 cup (85 grams) rolled oats
- 1 cup (225 grams) plain yogurt
- 1 cup (240 milliliters) whole milk
- ¼ cup (60 milliliters) vegetable oil
- ¼ cup (71 grams) molasses
- 1 large egg, at room temperature
- 1⅓ cups (185 grams) whole wheat flour
- 1 cup (142 grams) all-purpose flour
- 2 tablespoons dark brown sugar
- 2¼ teaspoons baking powder
- ¼ teaspoon baking soda
- 1¼ teaspoons salt
- 1 cup (142 grams) raisins
- ½ cup (71 grams) dried apricots, chopped

1 Preheat the oven to 375°F (190°C). Generously butter a 9 x 5-inch loaf pan and line it with parchment paper.

2 In a medium bowl, mix together the oats, yogurt, milk, oil, molasses, and egg until well blended. Set aside for at least 45 minutes or up to 90 minutes to allow the oats to hydrate.

3 In a large bowl, whisk together the whole wheat flour, all-purpose flour, brown sugar, baking powder, baking soda, and salt.

4 Gently stir the dry ingredients into the wet ingredients until just combined, then fold in the raisins and apricots.

5 Pour the batter into the prepared pan, spreading it evenly to the edges.

6 Bake the bread for 50 to 60 minutes, until a skewer inserted into the center comes out clean.

7 Turn out the loaf onto a wire rack and let cool for at least 1 hour before cutting and serving. Store the bread in an airtight container at room temperature for up to 3 days or keep refrigerated for up to 1 week. This bread also freezes really well and makes AMAZING toast.

CARROT and MARMALADE BREAD

Experimenting with flavors and taking risks is what makes a good baker a Bold Baker! I've known for a long time that carrots and oranges are a match made in heaven when it comes to flavor, but it was only recently that I created a recipe that let them both be the star of a recipe. You simply can't make vegetables taste any better than they do in this bread!

MAKES ONE 9 X 5-INCH LOAF

- 2 cups (284 grams) all-purpose flour
- 1 teaspoon ground cinnamon
- 1 teaspoon baking powder
- ½ teaspoon baking soda
- ½ teaspoon salt
- 2½ cups (282 grams) grated carrots (roughly 3 medium)
- ¾ cup (213 grams) bitter orange marmalade
- ¾ cup (1½ sticks/170 grams) butter, melted
- ½ cup (115 grams) granulated sugar
- 2 large eggs, at room temperature

FOR THE CREAM CHEESE GLAZE

- ¼ cup (57 grams) cream cheese, softened
- ½ tablespoon (7 grams) butter, softened
- ½ cup (57 grams) confectioners' sugar, sifted
- 1 tablespoon whole milk

1 Preheat the oven to 350°F (180°C). Butter a 9 x 5-inch loaf pan and line it with parchment paper.

2 In a large bowl, whisk together the flour, cinnamon, baking powder, baking soda, and salt.

3 In another large bowl, stir together the carrots, marmalade, melted butter, granulated sugar, and eggs.

4 Add the wet ingredients to the dry ingredients and stir until just combined. Pour into the prepared loaf pan.

5 Bake for 65 to 75 minutes, until a skewer inserted into the center comes out clean. Let the cake cool in the pan for 15 minutes.

6 **Make the cream cheese glaze:** In a small bowl, whisk together the cream cheese, butter, confectioners' sugar, and milk until smooth. Set aside.

7 Set a wire rack over a baking sheet, then turn the cake out onto the rack. Spread the cream cheese glaze over the cake, allowing it to drip all over.

8 Slice and enjoy with a cup of tea. Store leftovers in an airtight container at room temperature for up to 3 days.

Chapter 2

AFTERNOON TEA

I adore the tradition of afternoon tea. In fact, I love any tradition that brings people together to share good food. The beauty of these recipes—all designed for a light midday treat—is that you can share them with just one other person or a whole party. No matter who is sitting around your table spilling the tea (and eating cake), it's time well spent!

MIA'S BUTTERY SHORTBREAD COOKIES

Mia is a very important part of our team here at Bigger Bolder Baking because she cares for George while my husband, Kevin, and I are working. Mia is also from Ireland, and like any typical Irish mum, she never shows up to our home empty-handed. Be it toys or food, she takes good care of all of us.

These gorgeous, buttery shortbread cookies are one of Mia's specialties, which she so generously allowed me to include in this book. We can't state enough how grateful we are to her—and certainly not just for these cookies! George lights up when he sees Mia, and we know she'll always be part of our family.

MAKES 22 COOKIES

- 1 cup (2 sticks/225 grams) butter, softened
- ½ cup (57 grams) confectioners' sugar
- 2 teaspoons pure vanilla extract
- 2 cups (284 grams) all-purpose flour
- ¼ cup (28 grams) cornstarch
- ¼ teaspoon salt

1 In a large bowl, cream the butter, sugar, and vanilla together until light and fluffy.

2 In a small bowl, combine the flour, cornstarch, and salt. Add the dry ingredients to the butter mixture and stir until combined.

3 Turn the dough out of the bowl and shape it into a flattened disc. Wrap in plastic wrap and refrigerate for 1 hour, or until firm.

4 Preheat the oven to 325°F (170°C). Line two cookie sheets with parchment paper.

5 On a floured surface, roll out the chilled dough to ¼ inch thick. Using a 2-inch round cookie cutter, cut out as many cookies as you can, placing them 1 inch apart on the prepared baking sheets as you go. If you like, gather the scraps, roll them out, and cut more cookies.

6 Bake for 20 to 25 minutes, until the cookies are lightly golden around the edges. Transfer the cookies to wire racks and let cool completely.

7 Store the cooled cookies in an airtight container at room temperature for up to 1 week.

AUTHENTIC LINZER COOKIES

Linzer cookies are a holiday staple in many homes and at many cookie swaps, but they're not just a Christmas favorite for me—they're on my list of All-Time Favorite Cookies Ever! With a dollop of raspberry jam peeking through the top of a delicate, nutty, buttery cookie sandwich, all covered in a dusting of confectioners' sugar—you just can't beat a Linzer. It checks every box on my list for the perfect cookie, from texture to flavor!

MAKES 18 SANDWICH COOKIES

¾ cup (1½ sticks/170 grams) butter, softened

¾ cup (170 grams) granulated sugar

1 large egg, at room temperature

1 teaspoon pure vanilla extract

1½ cups (213 grams) all-purpose flour

½ cup (57 grams) almond flour

1 teaspoon ground cinnamon

½ teaspoon baking powder

½ teaspoon salt

½ cup (142 grams) seedless raspberry jam

Confectioners' sugar, for dusting

1 In a large bowl, cream the butter and granulated sugar together until pale and fluffy.

2 Add the egg and vanilla and beat until well combined.

3 In a medium bowl, whisk together the all-purpose flour, almond flour, cinnamon, baking powder, and salt. Gradually mix the flour mixture into the butter mixture until just combined.

4 Turn the dough out of the bowl and shape it into a flattened disc. Wrap in plastic wrap and refrigerate until very firm, at least 2 hours or up to 3 days.

5 When you are ready to bake your cookies, preheat the oven to 375°F (190°C). Line two cookie sheets with parchment paper.

6 On a floured surface, roll out the dough to about ⅛ inch thick. Using a 2½-inch round cookie cutter, cut out as many cookies as you can, placing them 2 inches apart on the prepared cookie sheets. If you like, gather the scraps, roll them out, and cut more cookies.

7 Using a 1-inch round cookie cutter, cut a hole in the center of half the cookies. (You can bake them up separately if you'd like.)

8 Bake for 15 to 18 minutes, rotating the sheets halfway through, until the cookie edges are lightly golden brown. Let cool on the cookie sheets for a few minutes before transferring to a wire rack to cool completely.

9 Flip over the cookies without holes so they are flat-side up and spread about 1 teaspoon of jam into the center of each, leaving a small border.

10 Dust the cut-out cookies generously with confectioners' sugar, then carefully place them over the jam-covered cookies. Press to seal very gently so you don't break the top cookies.

11 The assembled cookies are best enjoyed the day they are made, but leftovers can be stored in an airtight container at room temperature for up to 3 days.

HOMEMADE CHOCOLATE-OAT TEA COOKIES

Teatime is biscuit time in Ireland, and we take that very seriously! We always make sure to have a tin of biscuits on hand to serve to anyone who might stop by.

These oat biscuits smeared with chocolate are crisp and buttery and scream to be dunked into a hot cup of tea. Even though I live in California now, I still enjoy my teatime every day, with a bickie on the side, of course!

MAKES 16 COOKIES

FOR THE COOKIES

8½ tablespoons (125 grams) butter, at room temperature

⅓ cup (80 grams) dark brown sugar

2 tablespoons honey

⅔ cup (57 grams) rolled oats

⅔ cup (100 grams) whole wheat flour

½ teaspoon baking powder

½ teaspoon baking soda

½ teaspoon salt

FOR THE CHOCOLATE TOPPING

⅔ cup (3½ ounces/ 100 grams) finely chopped milk chocolate

2 tablespoons (30 grams) butter, softened

1 Preheat the oven to 350°F (180°C). Line a few cookie sheets with parchment paper.

2 **Make the cookies:** In a large bowl, cream the butter and brown sugar until pale and fluffy. Add the honey and beat until incorporated.

3 In a medium bowl, combine the oats, flour, baking powder, baking soda, and salt, then add the dry ingredients to the butter mixture and mix until just combined.

4 Cover the dough with plastic wrap and refrigerate for 15 to 20 minutes for easier handling.

5 Divide the chilled dough into 16 balls, placing them a few inches apart on the prepared cookie sheets, then flatten the balls into 2-inch discs.

6 Bake for 10 to 12 minutes, rotating the sheets halfway through, until golden brown. Let cool completely on the cookie sheets.

7 **Make the chocolate topping:** Fill a small saucepan with an inch or two of water and bring to a simmer. Put the chocolate and butter in a heatproof bowl and set it over the simmering water to melt (be sure the bottom of the bowl does not touch the water).

8 Spoon a teaspoon of the chocolate topping onto each cookie and spread it evenly using the back of the spoon. Let the topping set completely before enjoying. Store leftovers in an airtight container at room temperature for up to 3 days.

CLASSIC CHOCOLATE ÉCLAIRS

As a pastry chef, I'm no stranger to a classic chocolate éclair. During culinary school and throughout my professional career, I have made my fair share of éclairs! They're elegant, they're fancy, and they make any occasion just that much more special.

They're also not that difficult to make! *Éclair* means "flash of lightning" in French because they're eaten in a *flash*. And the choux pastry, the base of a chocolate éclair, can also be made in a flash!

MAKES 18 ÉCLAIRS

FOR THE CHOUX PASTRY

- 1 cup (240 milliliters) water
- ½ cup (1 stick/142 grams) butter, softened
- 1 tablespoon granulated sugar
- ½ teaspoon salt
- 1 cup (142 grams) all-purpose flour
- 3 large eggs, at room temperature

FOR FILLING AND TOPPING

- 1 recipe Crème Chantilly (see page 232)
- 1 recipe Chocolate-Butter Glaze (page 237), pouring consistency

1 Preheat the oven to 400°F (200°C). Line two cookie sheets with parchment paper.

2 **Make the choux pastry:** In a medium saucepan, combine the water, butter, sugar, and salt and bring to a boil over medium heat. When the butter has melted and the mixture reaches a full boil, remove the saucepan from the heat and stir in the flour all at once until the mixture begins to pull away from the sides of the pan.

3 Return the saucepan to the stove over low heat and cook, continuously smearing the dough against the pan, for about 3 minutes, until the mixture is slightly shiny.

4 Immediately transfer the mixture to a bowl and whisk for about 20 seconds to cool slightly.

5 Add the eggs to the batter one at a time, whisking until incorporated after each addition, until a smooth, thick, sticky paste forms.

6 Transfer the batter to a piping bag fitted with a large open star tip and pipe thick strips of batter, 5 inches long and a few inches apart, onto the prepared baking sheets.

7 Bake for 20 to 25 minutes, rotating the pans halfway through, until puffed and golden brown. Turn the oven off, crack open the oven door, and let cool completely while still in the oven (this will help keep their shape).

8 Split the cooled pastries almost in half lengthwise and generously pipe or spread with the crème chantilly.

9 Dip the tops of the éclairs into the warm glaze and let set before serving.

10 These are best eaten the day they are made, but leftover éclairs can be stored in an airtight container in the fridge for up to 2 days.

CAPPUCCINO SWISS ROLL

Swiss rolls are synonymous with teatime in Ireland, but I wanted to take a bold twist on the usual sponge cake with its pinwheel of jam. Instead, in this recipe I've paired dark chocolate and coffee to make a showstopping dessert! It has a much more refined taste than your typical Swiss roll.

MAKES 8 SERVINGS

FOR THE CAKE

- 4 large eggs, at room temperature
- ½ cup (115 grams) granulated sugar
- 2 tablespoons vegetable oil
- 1 teaspoon pure vanilla extract
- ½ cup (71 grams) all-purpose flour
- ¼ cup (28 grams) unsweetened cocoa powder
- 1 teaspoon baking powder
- 1 teaspoon instant espresso powder
- ½ teaspoon salt

FOR THE FILLING AND TOPPING

- 1 recipe Coffee Cream (see page 232)
- 1 recipe Chocolate-Butter Glaze (page 237), pouring consistency

1 **Make the cake:** Preheat the oven to 350°F (180°C). Butter a 15 x 10-inch jelly-roll pan and line it with parchment paper, leaving a slight overhang.

2 In the bowl of a stand mixer fitted with the whisk attachment or in a large bowl using a handheld mixer, beat the eggs , sugar, oil, and vanilla together until smooth and thick.

3 Place a sieve over the bowl and sift in the flour, cocoa powder, baking powder, instant espresso, and salt. Fold in the dry ingredients until just combined.

4 With a spatula, spread the batter evenly into the prepared pan.

5 Bake for about 11 minutes, until the top of the cake springs back when touched. Set aside to cool slightly.

6 While the cake is still warm, use the parchment to lift it out onto your work surface. Starting at one 10-inch (shorter) end, roll up the cake with the parchment and let cool completely. (This step will make the cake easier to roll—and hopefully help keep it from cracking—after you've filled it later on.)

7 Once cool, gently unroll the cake. Spread the coffee filling evenly over its surface, leaving a 1-inch border all around.

8 Starting with the short end, gently begin rolling the cake in a tight roll. Use the parchment paper as a guide to help you roll.

9 Place the rolled cake seam-side down on a serving tray. Drizzle with the glaze and let it set before slicing and serving. Store leftovers, covered, in the fridge for up to 2 days.

CHOCOLATE SWIRL MERINGUES

When it comes to choosing which desserts are on my list of favorites, texture plays a huge part. A meringue has everything I love! Every single part of a meringue has a different texture; it's crispy on the outside, chewy and marshmallow-y in the middle, and, in this version, sandwiched with soft, freshly whipped cream. Add some chocolate to the mix, and that's my idea of heaven on earth!

MAKES 26 MERINGUES

1 recipe Basic Meringue (page 220)

⅔ cup (4 ounces/115 grams) chopped bittersweet chocolate, melted

1 recipe Whipped Cream (page 232)

1 Preheat the oven to 225°F (110°C). Line two cookie sheets with parchment paper.

2 Make the meringue. While it's still in the mixing bowl, drizzle the melted chocolate over the surface of the meringue. Without fully stirring in the chocolate, use two dessert spoons to drop 26 equal-size mounds of meringue onto the prepared cookie sheets, spacing them apart evenly.

3 Bake for about 1 hour 15 minutes, then turn off the oven. Leave the meringues in the oven until the oven is cold to help dry them. (The unfilled meringues can be stored in an airtight container at room temperature for up to 3 days.)

4 Just before serving, spread the whipped cream over the flat side of half the meringues and sandwich with the remaining meringues. Enjoy straight away.

MOIST FRUIT-and-NUT CAKE

Move over, sticky toffee pudding—there's a new date recipe in town! I love baking with dates; they're sweet but not *too* sweet, and they add a delicious moistness to this loaf. The nuts give it a great texture, making every bite exciting. This humble-looking quick bread is the perfect accompaniment to a cup of tea on a lazy afternoon. You'll never underestimate a delicious date-and-nut loaf again!

MAKES ONE 9-INCH CAKE (10 SERVINGS)

- 2 cups (284 grams) pitted dates, chopped
- 1 teaspoon baking soda
- ¾ cup (180 milliliters) boiling water
- 2 medium Granny Smith apples (about 10 ounces/284 grams), peeled, cored, and grated
- 10 tablespoons (1¼ sticks/142 grams) butter, melted
- ½ cup (85 grams) dark brown sugar
- 4 large eggs, at room temperature
- 1 tablespoon pure vanilla extract
- 1¼ cups (177 grams) all-purpose flour
- 2½ teaspoons baking powder
- ½ teaspoon salt
- ½ cup (71 grams) walnuts, toasted and finely chopped

1 Preheat the oven to 325°F (170°C). Butter a 9-inch round cake pan and line it with parchment paper.

2 Place the dates and baking soda in a small bowl and pour in the boiling water. Cover with plastic wrap and allow the dates to soften for 20 minutes.

3 Transfer the dates and soaking water to a food processor and process until smooth. Add the apples, butter, brown sugar, eggs, and vanilla and pulse until evenly mixed.

4 In a small bowl, whisk together the flour, baking powder, and salt, then add this to the food processor and pulse to combine. Use a spatula to stir in the walnuts.

5 Spread the batter into your prepared pan.

6 Bake for 50 to 60 minutes, until a toothpick inserted into the center comes out clean. Let cool in the pan for 20 minutes before serving.

7 Slice and enjoy with a cup of tea. Store leftovers in an airtight container at room temperature for up to 4 days.

MUM'S IRISH APPLE CAKE

My mum's apple crumb cake is one of her best (and my favorite) Irish recipes, and I'm so happy to share it with you—I know it will be an instant hit! A classic Irish apple crumb cake looks rustic and straightforward, but it is a diamond in the rough. It's somewhat of a cross between an American coffee cake and an apple pie. Can anyone name a better combination?

MAKES ONE 9-INCH CAKE (8 SERVINGS)

FOR THE STREUSEL TOPPING

- ¾ cup (105 grams) all-purpose flour
- ¼ cup (21 grams) rolled oats
- ½ cup (115 grams) granulated sugar
- ⅛ teaspoon salt
- 6 tablespoons (¾ stick/ 85 grams) cold butter, diced

FOR THE CAKE

- ½ cup (1 stick/115 grams) butter, softened
- ½ cup (115 grams) granulated sugar
- 3 tablespoons whole milk, at room temperature
- 2 teaspoons pure vanilla extract
- 2 large eggs, at room temperature
- 1¼ cups (177 grams) all-purpose flour
- 1 teaspoon baking powder
- 1 teaspoon ground cinnamon
- ⅛ teaspoon salt
- 3 medium Granny Smith apples (about 15 ounces/425 grams), peeled, cored, and thinly sliced

1 Preheat the oven to 350°F (180°C). Butter a 9-inch round cake pan and line it with parchment paper.

2 **Make the streusel topping:** In a medium bowl, combine the flour, oats, sugar, and salt.

3 Add the butter and, using your fingertips, rub it into the dry ingredients until the mixture resembles coarse bread crumbs. Refrigerate while you make the cake batter.

4 **Make the cake:** In a large bowl, cream the butter and sugar together until light and fluffy.

5 Add the milk and vanilla, then beat in the eggs one at a time.

6 In a medium bowl, combine the flour, baking powder, cinnamon, and salt.

7 Fold the dry ingredients into the wet ingredients until evenly mixed.

8 Spread the batter into the prepared cake pan. Arrange the apple slices evenly over the batter, then crumble the streusel topping over the apples.

9 Bake for 60 to 70 minutes, until the top is golden brown all over and crisp. Let cool in the pan for 15 minutes, then turn the cake out onto a wire rack to cool completely.

10 When ready to serve, slice and enjoy. Store leftovers in an airtight container at room temperature for up to 3 days.

CITRUS OLIVE OIL POUND CAKE

I'm late to the party on baking with olive oil, and more fool me for not adding it to my desserts sooner! Olive oil doesn't just add moisture; it also adds a lovely nuttiness to your baked goods. You will definitely taste the difference when you bake with olive oil. It has such a beautifully unique flavor.

I love pairing a good, dense pound cake with a light, zesty flavor like citrus; I find that it perfectly rounds out the richness of the olive oil cake.

MAKES 12 SERVINGS

5 large eggs, at room temperature

3 cups (675 grams) granulated sugar

1½ cups (340 milliliters) extra-virgin olive oil

1½ cups (340 milliliters) fresh orange juice

Grated zest of 2 lemons

3½ cups (497 grams) all-purpose flour

1½ teaspoons baking powder

1¾ teaspoons salt

Confectioners' sugar, for dusting

1 Preheat the oven to 350°F (180°C). Generously butter a large (15-cup/3½-liter) Bundt pan and dust it with flour, tapping out any excess.

2 In the bowl of a stand mixer fitted with the whisk attachment or in a large bowl using a handheld mixer, beat the eggs and granulated sugar on high for about 3 minutes, until pale and fluffy.

3 Add the olive oil, orange juice, and lemon zest and beat until fully combined.

4 In a large bowl, whisk together the flour, baking powder, and salt, then add the dry ingredients to the wet ingredients in three additions, mixing well after each, until evenly incorporated.

5 Pour the batter into the prepared pan.

6 Bake for 65 to 75 minutes, until a wooden skewer inserted into the center comes out clean. If the top starts to brown before the cake is done, cover it loosely with aluminum foil. Let cool in the pan for 20 minutes, then turn the cake out onto a wire rack to cool completely.

7 Just before serving, dust the cake with confectioners' sugar. Store leftovers in an airtight container at room temperature for up to 5 days.

ALL-AMERICAN COFFEE CRUMB CAKE

Coffee cake is an oldie but a goodie, and I think every Bold Baker needs this cinnamon coffee crumb cake in their repertoire.

Technically, any cake you plan to eat with a cup of coffee is considered a coffee cake, but not in my book! (Literally.) This crumb cake is what I think of when I think of a classic coffee cake. Cut yourself a big wedge, pour a strong cup of coffee, relax, and repeat!

MAKES 12 SERVINGS

FOR THE STREUSEL LAYER

1¾ cups (247 grams) all-purpose flour

1 cup (170 grams) dark brown sugar

1¼ teaspoons ground cinnamon

½ teaspoon salt

¾ cup (1½ sticks/170 grams) cold butter, diced

1½ cups (213 grams) pecans, toasted and chopped

FOR THE CAKE

½ cup (1 stick/115 grams) butter, softened

1 cup (225 grams) granulated sugar

2 large eggs, at room temperature

1 cup (225 grams) sour cream

2 cups (284 grams) all-purpose flour

1¼ teaspoons baking powder

½ teaspoon baking soda

½ teaspoon salt

FOR THE MAPLE GLAZE

½ cup (57 grams) confectioners' sugar

1 tablespoon pure maple syrup

1 tablespoon whole milk

1. Preheat the oven to 350°F (180°C). Butter an angel food cake pan and dust it with flour, tapping out any excess. (If you don't have an angel food cake pan, you can use a Bundt pan or a 10-inch round cake pan.)

2. **Make the streusel layer:** In a medium bowl, stir together the flour, brown sugar, cinnamon, and salt.

3. Add the butter and, using your fingertips, rub it into the dry ingredients until the mixture resembles coarse bread crumbs. Stir in the pecans and set aside while you prepare the cake batter.

4. **Make the cake:** In the bowl of a stand mixer fitted with the whisk attachment, cream the butter and sugar together until pale and fluffy.

5. Add the eggs one at a time, followed by the sour cream.

6. In a medium bowl, whisk together the flour, baking powder, baking soda, and salt, then add the dry ingredients to the wet ingredients and mix until combined.

7. Spoon half the batter into the prepared pan. Sprinkle half the streusel mixture evenly over the batter, then top with the remaining batter and spread it evenly using a spatula. Sprinkle the remaining streusel evenly over the top.

8. Bake for 55 to 65 minutes, until the cake is golden brown and a toothpick inserted into the center comes out clean. Let cool in the pan for 10 minutes, then turn the cake out onto a wire rack.

9. **Make the maple glaze:** In a small bowl, whisk together the confectioners' sugar, maple syrup, and milk until smooth and shiny. Drizzle the glaze over the warm cake.

10. Cut a nice big slice for yourself and enjoy. Once the cake has cooled completely, it can be stored in an airtight container at room temperature for up to 3 days.

FRENCH YOGURT POT CAKE

When it comes to my baking, simplicity is key. I love recipes that I can whip up without too much thought. If baking is easy, it's more fun! And if you love a recipe that isn't just easy to make but easy to clean up, too, go ahead and bookmark this page.

You can make this delicious cake by simply measuring all of your ingredients using one yogurt tub. Yes, you read that right, *one yogurt tub!* I swear, that is some culinary magic!

For this recipe, you need a 6-ounce (170-gram) tub of plain yogurt (that's roughly ¾ cup). After using it to measure the wet ingredients, rinse and dry the tub before measuring the dry ingredients.

MAKES ONE 9-INCH CAKE (10 SERVINGS)

1 (6-ounce/170-gram) tub plain yogurt

2 yogurt tubs granulated sugar (12 ounces/ 340 grams)

1 yogurt tub vegetable oil (6 fluid ounces/ 180 milliliters)

3 large eggs, at room temperature

1 teaspoon pure vanilla extract

3 yogurt tubs all-purpose flour (11½ ounces/315 grams)

1 tablespoon baking powder

½ teaspoon salt

½ cup (5 ounces/142 grams) raspberry jam, for filling

Confectioners' sugar, for dusting

1 Preheat the oven to 350°F (180°C). Butter a 9-inch springform pan and line it with parchment paper.

2 In a large bowl, whisk together the yogurt, granulated sugar, and oil. Add the eggs and vanilla and whisk until combined.

3 Rinse and dry the yogurt pot, and then add the flour, baking powder, and salt to the batter and fold in until evenly mixed. Spread into the prepared pan.

4 Bake for about 35 minutes, until a toothpick inserted into the center comes out clean. Let the cake cool in the pan for 30 minutes, then remove it from the pan and let cool completely on a wire rack.

5 Cut the cooled cake in half horizontally to make two layers. Spread the bottom layer with the jam, replace the top layer, and dust with confectioners' sugar. Store the cake in an airtight container at room temperature for up to 3 days.

AMI'S DOUBLE CHOCOLATE ALMOND COOKIES

I am very fortunate to work with people who are experts in their field, and Ami is one of those fantastic people. She makes me bake outside my comfort zone, and she has never steered me wrong. When Ami mentioned baking a chocolate cookie using almond flour, I said, "Go for it!"

With Ami, I knew this cookie would be a hit! We all loved these cookies so much, I had to share this recipe with you.

MAKES 25 COOKIES

- ½ cup (1 stick/115 grams) butter, softened
- ½ cup (85 grams) dark brown sugar
- ¼ cup (57 grams) granulated sugar
- 1 large egg, at room temperature
- 1 teaspoon pure vanilla extract
- 2½ cups (287 grams) almond flour
- ½ cup (57 grams) unsweetened cocoa powder, sifted
- ½ teaspoon baking soda
- ½ teaspoon salt
- ¾ cup (3 ounces/85 grams) chopped bittersweet chocolate

1 Preheat the oven to 375°F (190°C). Line two cookie sheets with parchment paper.

2 In the bowl of a stand mixer fitted with the paddle attachment or in a large bowl using a handheld mixer, cream the butter, brown sugar, and granulated sugar on high speed until light and fluffy.

3 Add the egg and vanilla and beat until incorporated.

4 In a medium bowl, combine the almond flour, cocoa powder, baking soda, and salt. Add the dry ingredients to the wet ingredients and mix until incorporated, then fold in the chocolate by hand.

5 Roll 2-tablespoon scoops of dough into balls, placing them on the prepared cookie sheets as you go, then flatten the balls to about ½ inch thick.

6 Bake for about 8 minutes, until the cookies are puffed and a little cracked.

7 Let cool on the cookie sheets for about 5 minutes, then transfer to a wire rack to cool completely. Store in an airtight container at room temperature for up to 4 days.

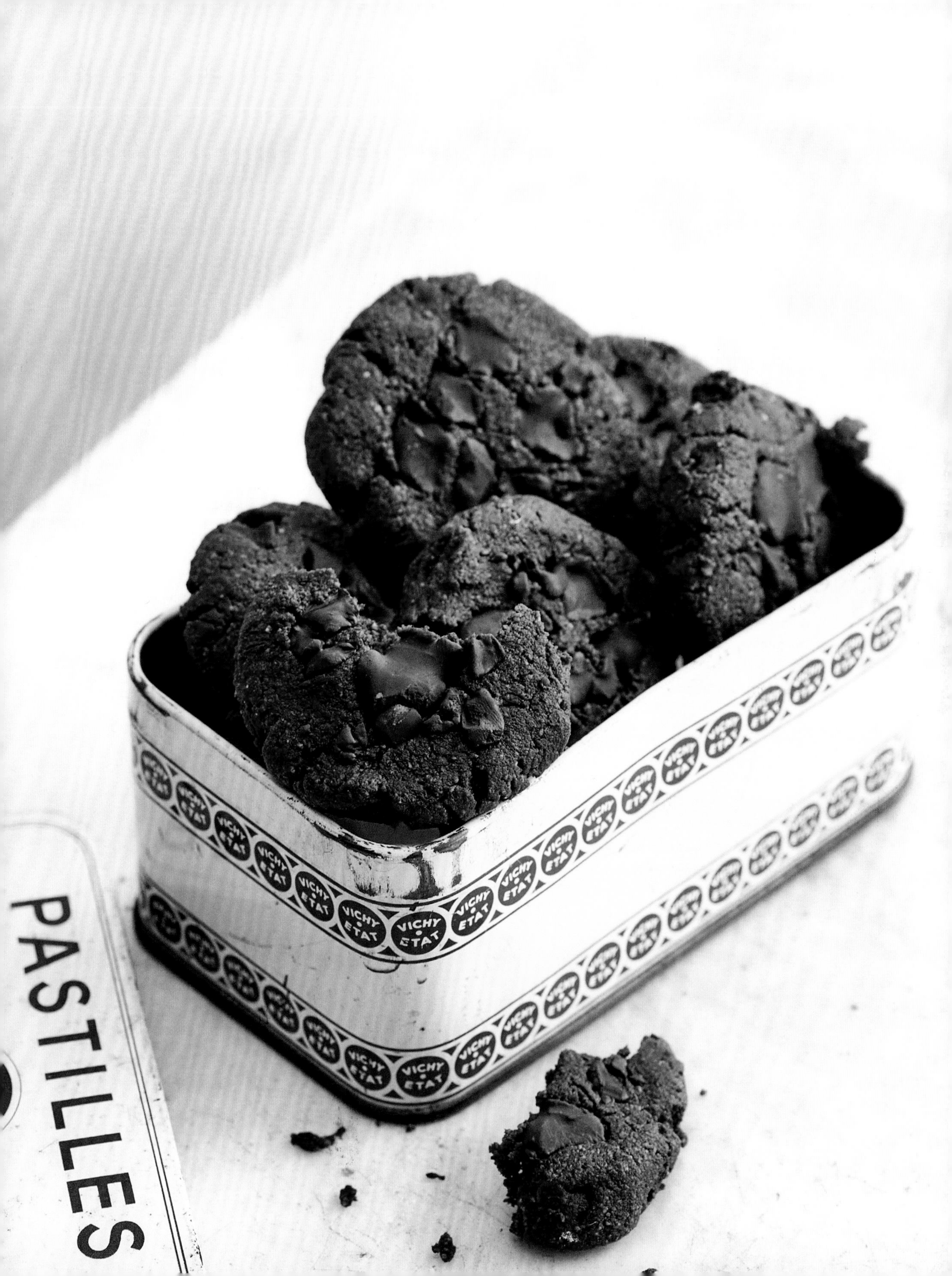
VICHY
ETAT
PASTILLES

BUTTERSCOTCH BANANA CAKE

There's more than one way to skin a banana! In the case of this banana bread, I add a lot of extra flavor by caramelizing the bananas before adding them to the rest of the ingredients. The result is out of this world! I guarantee, once you try this technique, you'll never look back!

MAKES ONE 8-INCH CAKE
(9 SERVINGS)

⅔ cup (142 grams) plus ½ cup (115 grams) granulated sugar

2 tablespoons water

3 very ripe small bananas (about 300 grams), peeled and mashed

1 tablespoon (28 grams) butter

¾ cup (180 milliliters) vegetable oil

¼ cup (57 grams) plain yogurt

2 large eggs, at room temperature

2 teaspoons pure vanilla extract

1¼ cups (177 grams) all-purpose flour

½ cup (71 grams) whole wheat flour

2 teaspoons baking powder

2 teaspoons ground cinnamon

½ teaspoon salt

Honey Toffee Sauce (page 230), for serving

1 Preheat the oven to 350°F (180°C). Butter an 8-inch square baking pan and line it with parchment paper.

2 In a medium saucepan, combine ⅔ cup (142 grams) of the sugar and the water and cook over medium heat until the sugar melts and turns a deep amber color, 8 to 10 minutes. (The deep color will give you more flavor, so don't be afraid to go dark with it—just don't let it burn!)

3 Stir in the mashed bananas and butter and cook for 2 minutes, or until the mixture is thick. Transfer to a large bowl and let cool for 10 minutes.

4 Add the remaining ½ cup (115 grams) sugar, the oil, yogurt, eggs, and vanilla and mix until blended.

5 In a small bowl, whisk together the all-purpose flour, whole wheat flour, baking powder, cinnamon, and salt. Fold into the banana mixture until just combined. Spread the batter into the prepared pan.

6 Bake for about 50 minutes, until a toothpick inserted into the center comes out clean. Let cool in the pan for 10 minutes, then turn the cake out onto a wire rack to cool completely.

7 Transfer the cake to a serving plate, pour over the honey toffee sauce, and enjoy! Store leftovers in an airtight container at room temperature for up to 3 days.

POLENTA CAKE *with* MASCARPONE *and* STRAWBERRY COMPOTE

It is such a treat to be served this beautiful cake for afternoon tea! I love the cake's soft, tender crumb, which is topped with fluffy sweetened mascarpone and a juicy strawberry compote.

As they say in Ireland, "There is eating and drinking in it!"

MAKES ONE 9-INCH CAKE (8 SERVINGS)

FOR THE CAKE

¾ cup (180 milliliters) buttermilk

½ cup (85 grams) instant polenta

1½ cups (213 grams) all-purpose flour

1½ teaspoons baking powder

¾ teaspoon baking soda

½ teaspoon salt

¾ cup (1½ sticks/170 grams) butter, softened

¾ cup (170 grams) granulated sugar

2 large eggs, at room temperature

FOR THE TOPPING

1 cup (225 grams) mascarpone cheese

¼ cup (58 grams) granulated sugar

⅓ cup (71 milliliters) heavy cream

1 recipe Strawberry Compote (page 234)

1 **Make the cake:** In a small bowl, combine the buttermilk and the polenta. Set aside to soak and hydrate for 45 minutes.

2 Preheat the oven to 350°F (180°C). Butter a 9-inch round cake pan and line it with parchment paper.

3 In a large bowl, whisk together the flour, baking powder, baking soda, and salt.

4 In a separate large bowl using a handheld mixer, cream the butter and sugar together until fluffy. Whisk in the eggs, one at a time, followed by the polenta mixture.

5 Fold in the flour mixture in three increments until just fully combined, taking care not to overmix. Pour the batter into your prepared pan.

6 Bake for 35 to 40 minutes, until a toothpick inserted into the center comes out clean. Run a thin knife around the edge of the pan to loosen the cake, then let the cake cool in the pan for 10 minutes before inverting it onto a wire rack to cool completely.

7 **Make the topping:** In a medium bowl, whisk together the mascarpone and granulated sugar, then add the cream and whisk to soft peaks.

8 Once the cake has cooled completely, transfer it to a serving plate. Top with the sweetened mascarpone and then the strawberry compote. Slice and enjoy! Store leftovers in an airtight container in the fridge for up to 2 days.

Chapter 3

WEEKNIGHT FAMILY FAVORITES

Some of my fondest memories from growing up in Ireland include my big family gathered around the table for the meals my mum would make. I vividly remember cold, dark winter nights spent watching the bread pudding bake or the times I helped my mum make the topping for apple crumble (still one of my favorites to this day!).

I cherish those memories, and now I get to give them to my son, George, as well. I hope once you try the recipes in this chapter, they'll become beautiful memories to share with your loved ones, too.

10-MINUTE SUMMER BERRY TIRAMISU

I love the traditional flavors and textures of an iconic Italian tiramisu, so I wanted to give my own version a go, and I'm really pleased with the result! This berry tiramisu has a thick, creamy, mousselike layer, delicious poached summer berries, and soft ladyfingers. You can't go wrong with that combination!

To say I have a sweet spot for this dessert (pun intended) may be an understatement. It has everything I love; plus, I can whip it up in just a few minutes! Dare I say it? If I could have only one dessert for the rest of my life, this would be a top contender.

MAKES 6 SERVINGS

- ¾ cup (170 grams) plus ⅓ cup (71 grams) granulated sugar
- ¾ cup (180 milliliters) water
- 1 cup (5 ounces/142 grams) fresh raspberries
- 1 cup (5 ounces/142 grams) fresh blueberries
- 2 cups (10 ounces/ 284 grams) fresh strawberries, quartered
- 1¼ cups (300 milliliters) heavy cream
- 1 teaspoon pure vanilla extract
- 1 cup (225 grams) mascarpone cheese
- 2 tablespoons crème de cassis (optional)
- 24 ladyfingers
- ¼ cup (28 grams) confectioners' sugar
- ⅓ cup (43 grams) slivered almonds, toasted

1 In a medium saucepan, combine ¾ cup (170 grams) of the sugar and the water. Heat over medium heat, stirring, until the sugar has dissolved, then turn off the heat and stir in the berries. Set aside to cool without stirring.

2 Strain the syrup from the berries into a medium bowl, reserving the berries in a separate bowl. If using, stir the crème de cassis into the strained syrup.

3 Using a stand mixer or an electric hand mixer, whip the cream, remaining sugar, and vanilla until soft peaks form. Add the mascarpone and mix until just combined.

4 **Assemble the tiramisu:** One at a time, briefly (for 3 or 4 seconds) dip the ladyfingers in the berry-infused syrup, then place in an 8-inch square baking dish, until you have covered the bottom of the dish to make the first layer (you can fit about 10 ladyfingers in each layer).

5 Spoon half the poached berries over the ladyfingers, then top with half the mascarpone mixture. Top with a second layer of syrup-dipped ladyfingers, the remaining berries, and finally the remaining mascarpone mixture. Refrigerate for at least 4 hours to set and to allow the flavors to develop.

6 Just before serving, sprinkle the tiramisu with the slivered almonds. Scoop into a bowl and enjoy! Store leftovers, covered, in the fridge for up to 1 day.

OLD-FASHIONED BANANA PUDDING

Banana pudding is not a dessert I grew up eating; thankfully, I'm making up for lost time now! Old-fashioned banana pudding is a Southern classic in America, and now it's a favorite around the country—and one of mine, too. It isn't a traditional Irish dessert, but I have learned a few tips: don't be shy with the salt, and don't be afraid to add vanilla to your pudding. That bit of advice will serve you well when it comes to a lot of other recipes, too!

MAKES 8 SERVINGS

2½   cups (600 milliliters) whole milk

2½   cups (600 milliliters) heavy cream

1   large egg, at room temperature

2   large egg yolks, at room temperature

¾   cup (170 grams) granulated sugar

¼   cup (30 grams) cornstarch

4   tablespoons (½ stick/ 57 grams) butter, diced and softened

1   tablespoon pure vanilla extract

⅛   teaspoon pure vanilla bean paste (optional)

½   teaspoon salt

5   to 6 cups (340 grams) lightly crushed Nilla Wafers

4   medium bananas (472 grams), peeled and sliced into ½-inch-thick rounds

1   medium banana, sliced, for decoratio

1   In a medium saucepan, combine the milk and ½ cup (120 milliliters) of the cream and bring to a simmer over medium heat.

2   In a medium bowl, whisk together the egg, egg yolks, sugar, and cornstarch until well blended.

3   While whisking continuously, ladle the hot milk mixture into the eggs, 1 cup at a time, to temper the eggs. Return the custard mixture to the saucepan and place it over low heat.

4   Cook the custard, whisking continuously, for 6 to 8 minutes, until it has thickened, then immediately pour the custard through a strainer into a large bowl to remove any curds that may have formed.

5   While the custard is still hot, whisk in the butter, vanilla extract, vanilla paste (if using), and salt. Set aside to cool completely (the custard will thicken as it cools).

6   Whip the remaining 2 cups (480 milliliters) of cream to soft peaks and refrigerate until ready to use.

7   **Assemble the pudding:** Cover the bottom of a trifle bowl with half the crushed Nilla Wafers. Spread half the custard on top, followed by half of the bananas. Dollop half the whipped cream on top and repeat the same layers once again: wafers, pudding, bananas, and cream. Cover and refrigerate for at least 4 hours or up to overnight to let the flavors develop before serving.

8   Just before serving, top with some sliced banana. Store leftovers, covered, in the fridge for up to 2 days.

STEAMED MARMALADE PUDDING

I grew up on puddings in Ireland, which are very different from puddings you'd likely find in the US. Instead of a dessert with a custardlike texture, puddings across the pond are actually cakes!

A popular method of cooking pudding is by using steam. If you're new to steamed puddings, hang on to your hat, because I'm about to rock your world! Metaphorically speaking, of course.

MAKES 8 SERVINGS

- 1 medium orange, peel on, coarsely chopped and seeded
- 1¼ cups (177 grams) all-purpose flour
- ⅔ cup (115 grams) light brown sugar
- ½ cup (1 stick/115 grams) butter, softened
- 2 large eggs, at room temperature
- 1 teaspoon ground ginger
- 1 teaspoon baking powder
- ½ teaspoon baking soda
- ½ teaspoon salt
- ¼ cup (115 grams) coarse-cut orange marmalade

1 Place a steamer basket in a large pot with a lid. Fill the pot with water to the bottom of the basket and place it on the stove over medium-low heat.

2 Butter the bottom and sides of a 1.4-liter pudding basin or a medium glass bowl. Set aside.

3 Put the orange in a food processor and process until finely chopped. Transfer to a small bowl.

4 In the food processor, combine the flour, brown sugar, butter, eggs, ginger, baking powder, baking soda, and salt and process until smooth and creamy, about 20 seconds. Add the chopped orange to the flour mixture and pulse a few times to combine.

5 Spoon the marmalade over the bottom of the prepared pudding basin, then top with the cake batter. Place a round of parchment paper on top of the batter and then cover with aluminum foil. Secure with a string around the basin to make it easy to lift from the pot.

6 Place the pudding basin in the steamer basket, cover the pot, and steam for about 2 hours, until the pudding is firm in the center and springs back when pressed. About every 20 minutes, carefully check the water level in the pot and add more hot water when necessary. Carefully remove the pudding from the steamer. While it is still warm, remove the parchment and foil and turn the pudding out onto a serving plate.

7 Serve the pudding warm, with custard or vanilla ice cream. Store leftovers, covered, at room temperature for up to 3 days.

FLAKY PEACH SHORTCAKES

I'm relatively new to the world of American shortcakes, but as a person who comes from the country of the scone, I feel like I am predisposed to love this sweet treat! My flaky peach shortcake is made with a buttery, sweet biscuit and macerated peaches, and it's topped with a big, fat dollop of whipped cream. I'm sold!

FOR THE BISCUITS

- 3 cups (426 grams) all-purpose flour
- 2 tablespoons granulated sugar
- 4 teaspoons baking powder
- ¼ teaspoon salt
- ¾ cup (1½ sticks/170 grams) cold butter, diced
- 1½ cups (360 milliliters) heavy cream
- 1 egg, beaten, for egg wash

FOR THE FILLING

- 6 cups pitted and sliced peaches
- ⅓ cup (71 grams) granulated sugar
- 1 teaspoon pure vanilla extract
- 1 teaspoon pure vanilla paste (optional)
- 1 recipe Whipped Cream (page 232), for topping

1 **Make the biscuits:** Preheat the oven to 400°F (200°C). Line a baking sheet with parchment paper.

2 In a large bowl, combine the flour, granulated sugar, baking powder, and salt.

3 Add the butter and, using your fingertips, rub it into the dry ingredients until the mixture resembles coarse bread crumbs. Pour in the cream and stir quickly until the dough comes together (if the dough is still a little dry, add a splash more liquid).

4 Turn the dough out onto a lightly floured surface, press it together to incorporate any loose dough, and roll it out to about 1 inch thick. Cut eight 3-inch rounds or squares from the dough. Any scraps can be pressed together, rerolled, and cut into additional biscuits.

5 Place the biscuits on the prepared baking sheet and brush the tops with the egg wash.

6 Bake for 20 to 25 minutes, until golden brown. Let cool completely. (The cooled unfilled biscuits can be stored in an airtight container at room temperature for up to 3 days before assembly.)

7 **Make the filling:** In a medium bowl, combine the peaches, sugar, vanilla extract, and vanilla paste (if using). Set aside for 20 minutes to macerate, stirring occasionally.

8 To assemble the shortcakes, gently split the biscuits in half. Spoon some of the peach mixture onto the bottom halves, add a dollop of the whipped cream, and sandwich with the biscuit tops. Serve immediately.

WHOLE LEMON TART

I love lemons, especially desserts made with lemons, but the only thing I hate as much as I love lemons is juicing them. I know *hate* seems like a very strong word, especially when used against juicing a lemon, but I stand by it!

I adore this whole lemon tart because it is bursting with citrus flavor, with very little juicing and no zesting required. You simply throw all your filling ingredients—including the lemon itself!—into a blender. The results speak for themselves!

MAKES ONE 9-INCH TART (8 SERVINGS)

- 1 recipe Pie Crust (page 222)
- 1 medium lemon (about 4½ ounces/130 grams), scrubbed
- 1½ cups (340 grams) granulated sugar
- ½ cup (1 stick/115 grams) butter, softened
- 4 large eggs, at room temperature
- 1 large egg yolk, at room temperature
- 3 tablespoons fresh lemon juice
- 2 tablespoons cornstarch
- ¼ teaspoon salt
  Whipped Cream (page 232), for serving

1 Preheat the oven to 350°F (180°C). On a floured surface, roll out the pie dough into a round about ⅛ inch thick. Transfer the dough to a 9-inch pie pan, then line it with parchment paper and fill with pie weights (or dried beans).

2 Blind-bake the crust for about 20 minutes, then remove the parchment paper and pie weights and bake for another 5 to 10 minutes to let the bottom bake.

3 Meanwhile, quarter and seed the lemon, leaving the skin on, then put the lemon quarters in a food processor. Add the sugar and butter and process until very smooth, 2 to 3 minutes. Add the eggs, egg yolk, lemon juice, cornstarch, and salt and process until combined.

4 Set the prebaked crust on a baking sheet and pour in the filling.

5 Bake for 35 to 40 minutes, until the filling is set but still jiggles slightly in the middle when tapped. Let cool for at least 2 hours before serving.

6 Slice and serve with freshly whipped cream. Store leftovers, covered, in the fridge for up to 2 days.

CHOCOLATE LOVER'S CHEESECAKE *with* STRAWBERRY COMPOTE

Warning! This decadent dessert takes *no* prisoners, and I cannot be held liable for how many times you end up remaking this incredibly rich in flavor and silky-smooth chocolate cheesecake. If chocolate is your vice, then you've come to the right dessert; go ahead and dog-ear this page.

MAKES ONE 9-INCH CHEESECAKE (10 SERVINGS)

- 27 Oreo cookies (300 grams)
- 2 cups (12 ounces/ 340 grams) chopped bittersweet chocolate
- 3 cups (680 grams) full-fat cream cheese, at room temperature
- 1 cup (225 grams) granulated sugar
- 3 large eggs, at room temperature
- 1 cup (225 grams) full-fat sour cream, at room temperature
- 1½ teaspoons pure vanilla extract
  Whipped Cream (page 232), for serving
- 1 recipe Strawberry Compote (page 234), for serving

1 Preheat the oven to 350°F (180°C). In a food processor, blend the cookies, including their filling, until fine and sticking together. Firmly press the crushed Oreos into an even layer over the bottom of a 9-inch springform pan. Set aside.

2 Place the chocolate in a medium microwave-safe bowl. Microwave in 30-second intervals, stirring after each, until melted and smooth, then set aside to cool slightly. (Alternatively, melt the chocolate using a double boiler.)

3 In the bowl of a stand mixer fitted with the paddle attachment or in a large bowl using a handheld mixer, beat the cream cheese and sugar together until smooth. Pour in the melted chocolate and mix until combined.

4 Add the eggs one at a time, beating each until incorporated before adding the next. Add the sour cream and vanilla and mix until evenly combined.

5 Pour the batter over the Oreo crust and smooth the top.

6 Bake for 50 to 60 minutes, until the cheesecake looks firm yet still jiggles a little in the middle. Remove from the oven and carefully run a knife or spatula around the inside edge of the pan to loosen the cheesecake (this helps prevent the surface from cracking as it cools). Let cool in the pan to room temperature, then cover and refrigerate for at least 4 hours or up to overnight.

7 When ready to serve, remove the springform ring, then slice and top with the whipped cream and strawberry compote. Store leftovers, covered, in the fridge for up to 3 days.

BAKED CUSTARD *with* RHUBARB COMPOTE

Delicious desserts don't have to be a daylong experience or made with exotic ingredients to be satisfying. Sometimes, the simplest recipes are exactly what you need, want, and *crave*! Case in point: this baked custard, sprinkled with nutmeg and the delicate flavors of rhubarb.

MAKES 6 SERVINGS

1⅔ cups (382 milliliters) whole milk

1½ cups (360 milliliters) heavy cream

1 vanilla bean

10 large egg yolks, at room temperature

¾ cup (170 grams) granulated sugar

Whole nutmeg, for grating

1 recipe Rhubarb Compote (page 236)

1 Preheat the oven to 275°F (140°C). Place an 8-inch square baking dish inside a 9 x 13-inch roasting pan. Place a strainer over the baking dish. Bring a kettle of water to a boil.

2 In a medium saucepan, combine the milk and cream. Split the vanilla bean in half, scrape out the seeds, and add the seeds and scraped pod to the milk mixture. Heat over medium heat until the mixture just comes to a simmer, then remove from the heat.

3 While the milk is heating, in a medium bowl, whisk together the egg yolks and sugar. While whisking continuously, slowly pour 1 cup of the milk mixture at a time into the egg mixture and whisk to combine.

4 Pour the milk-egg mixture through the strainer into the baking dish. Remove the strainer and grate nutmeg over the top of the custard.

5 Transfer the baking dish (still in the roasting pan) to the oven, then carefully pour boiling water from the kettle into the roasting pan until it reaches halfway up the sides of the baking dish.

6 Bake for 65 to 75 minutes, until the custard is set but still jiggles slightly in the middle when tapped. Carefully remove the roasting pan from the oven and let the custard cool slightly in the water, then remove the custard and refrigerate uncovered for about 4 hours, until set.

7 Spoon the custard into individual serving dishes and top with the compote. Store leftovers, covered, in the fridge for up to 3 days.

SMASHED RASPBERRY PAVLOVA

Some of my fondest food memories from growing up in Ireland include my mum's delicious pavlova. It's a dessert I still love and one that always reminds me of home and family.

This recipe is a fun one to share with the family. It looks incredibly impressive, but it's no fuss to make. My mum and my son were my inspirations for this smashed pavlova, and I hope that in years to come, George will tell his own children about his mum's amazing pavlova, too. It's a beautiful treat and one I hope you will enjoy, as well.

MAKES 8 SERVINGS

1 recipe Basic Meringue (page 220)
1 recipe Whipped Cream (page 232)
1 recipe Raspberry Vanilla Compote (page 236)
2 cups (10 ounces/ 283 grams) fresh raspberries
Fresh mint, for garnish

1 Preheat the oven to 250°F (130°C). Line a baking sheet with parchment paper.

2 Make the meringue, then spread it into a 10-inch round on the prepared baking sheet.

3 Bake for about 1½ hours, until the meringue feels dry to the touch and lifts easily off the parchment, then turn the oven off. Leave the meringue in the oven to cool for 2 hours. (The cooled meringue can be stored, covered, at room temperature for up to 2 days.)

4 When you are ready to serve the pavlova, place the meringue on a large serving platter and smash it into big pieces.

5 Dollop the whipped cream onto the crushed meringue, spoon the compote all over, and top with the fresh raspberries. Garnish with some mint and serve immediately. Store any leftovers, covered, in the fridge for up to 24 hours.

SQUIDGY CHOCOLATE CAKE

I'm forever trying to satisfy my chocolate craving—and when that craving hits particularly hard, I turn to this unbeatable chocolate mousse cake. You'd be hard-pressed to find another cake that is just *so* right for that urge. It's soft, bubbly, and *oh so* dreamy!

Serve this with some whipped cream on top, and I promise bowls will be licked clean!

**MAKES ONE 9-INCH CAKE
(8 SERVINGS)**

1   cup (6 ounces/170 grams) chopped bittersweet chocolate

¾   cup (180 milliliters) water

¾   cup (1½ sticks/170 grams) butter, softened

4   large eggs, separated, at room temperature

½   cup (115 grams) granulated sugar

1   teaspoon pure vanilla extract

¼   cup (35 grams) self-rising flour

1   recipe Whipped Cream (page 232), for serving

1   Preheat the oven to 400°F (200°C). Butter a 9-inch round baking dish and place it in a 9 x 13-inch roasting pan. Bring a kettle of water to a boil.

2   In a small saucepan, combine the chocolate and the water and heat over medium heat until the chocolate melts. (You can also do this in the microwave; combine the chocolate and water in a small microwave-safe bowl and heat in 30-second intervals, stirring after each, until melted.)

3   Transfer the chocolate to a large bowl and whisk in the butter, egg yolks, sugar, and vanilla until fully mixed. Stir in the flour.

4   In the bowl of a stand mixer fitted with the whisk attachment or in a medium bowl using a handheld mixer, whip the egg whites until they reach stiff peaks.

5   Carefully fold the egg whites into the chocolate mixture, then pour the batter into the prepared baking dish.

6   Transfer the baking dish (still in the roasting pan) to the oven, then pour boiling water from the kettle into the roasting pan until it reaches halfway up the sides of the baking dish.

7   Bake for 10 minutes, then reduce the oven temperature to 325°F (160°C) and bake for another 30 to 35 minutes, until the cake is firm around the edges but still gooey in the center; be careful not to overbake it, or you won't get that squidgy center. Let cool for a few minutes before serving.

8   Serve with whipped cream on the side. Store leftovers, covered, in the fridge for up to 3 days.

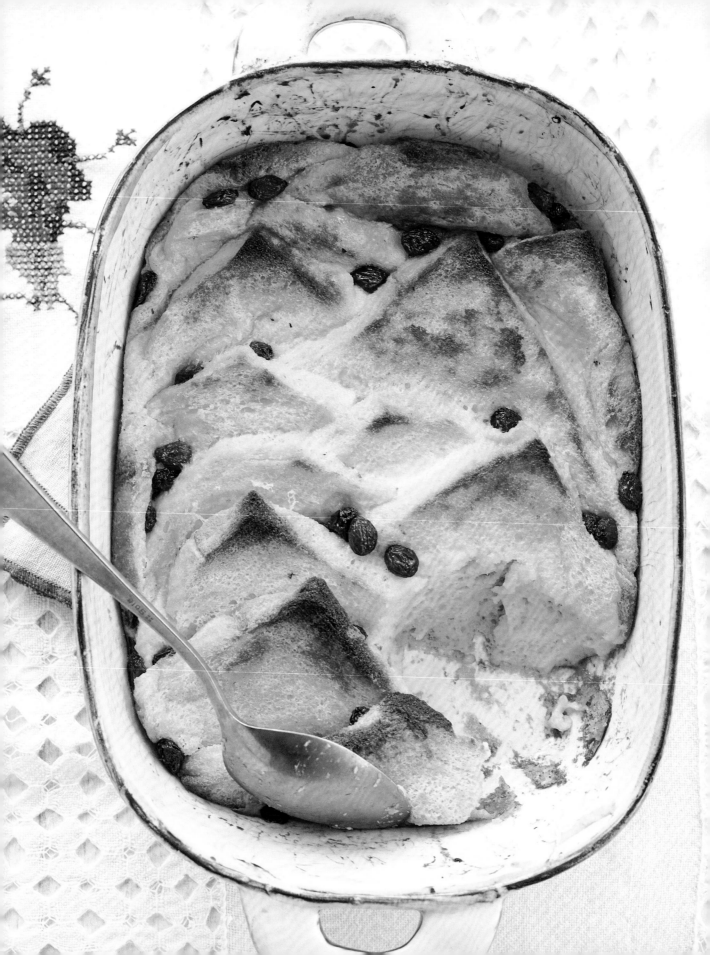

TRADITIONAL IRISH BREAD-*and*-BUTTER PUDDING

What type of Irish person would I be if I didn't include a traditional bread-and-butter pudding with my family favorites? I'm pretty positive it is required . . . it may have been written in the fine print on my birth certificate!

This recipe is an oldie but a goodie, and it holds many wonderful memories for me. It was one of my mum's go-to desserts on dark winter nights when I was growing up. Those memories have stayed with me all this time. It's a recipe I really wanted to share and one I hope becomes one of your family's favorite memories, too.

MAKES 6 SERVINGS

4 tablespoons (½ stick/ 57 grams) butter, softened

6 (½-inch-thick) slices white bread, crusts removed

¼ cup (1¼ ounces/ 45 grams) raisins

4 large eggs, at room temperature

2 cups (480 milliliters) whole milk

¼ cup (57 grams) granulated sugar

¼ teaspoon ground nutmeg
   Whipped Cream (page 232), for serving

1 Preheat the oven to 350°F (180°C) and butter an 11 x 8-inch (or similarly sized) baking dish.

2 Spread the butter on the bread slices, then cut them diagonally in half to make triangles. Lay the bread, buttered-side up and overlapping slightly, in the prepared baking dish. Scatter the raisins between the triangles of bread.

3 In a medium bowl, whisk together the eggs, milk, sugar, and nutmeg. Pour this mixture over the bread and set aside at room temperature for at least 45 minutes and up to 2 hours to let the bread absorb the liquid.

4 Bake for 50 to 60 minutes, until the bread pudding has risen and is golden in color.

5 Top with freshly whipped cream and serve immediately. Store leftovers, covered, in the fridge for up to 3 days.

CLASSIC BLUEBERRY PIE

They say you can't improve on perfection, and I think this classic blueberry pie is a testament to that rule. I love fresh summer blueberries, but when they're baked, *oh baby!*

Tucked into a flaky, delicious pie crust and baked, those delicious blueberries become an eye-catching violet color and take on a sweet, almost floral flavor. It's the quintessential summer pie!

MAKES ONE 9-INCH DOUBLE-CRUST PIE (8 SERVINGS)

2	recipes Pie Crust (page 222)
4	cups (20 ounces/ 568 grams) fresh blueberries, rinsed and dried
½	cup (115 grams) granulated sugar
3	tablespoons cornstarch
2	teaspoons grated lemon zest
1	tablespoon fresh lemon juice
¼	teaspoon salt
1	tablespoon (14 grams) butter, diced
1	egg, beaten, for egg wash
1	tablespoon coarse sugar
	Vanilla ice cream, for serving

1 Preheat the oven to 375°F (190°C). Line a baking sheet with parchment paper.

2 Roll out one disc of pie dough into a round roughly ⅛ inch thick, then transfer it to a 9-inch pie pan.

3 In a large bowl, combine the blueberries, granulated sugar, cornstarch, lemon zest, lemon juice, and salt. Pour the blueberry mixture into the pie crust and dot the top with the diced butter.

4 Roll out the remaining disc of pie dough into a roughly 10-inch round, about ⅛ inch thick.

5 Brush the rim of the filled pie crust with the egg wash and carefully lay the round of dough on top. Press the edges together with your fingers or a fork to seal.

6 Brush the top crust with egg wash and sprinkle with coarse sugar. Cut a few slits in the top crust to allow steam to escape.

7 Place the pie on the prepared baking sheet.

8 Bake for about 1 hour, until the juices start to bubble. Let the pie cool for at least 4 hours before slicing.

9 Slice and serve with a scoop of vanilla ice cream on top. Store leftover pie, covered, at room temperature for up to 3 days.

THE ULTIMATE WHITE CHOCOLATE PECAN SKILLET COOKIE

I don't use the term *ultimate* loosely; this dessert is made to take center stage, smack-dab in the middle of your table, where everyone can help themselves! You're guaranteed to see smiles when you serve this giant cookie that's perfectly crisp on the edges and gooey in the middle, chock-full of crunchy pecans and smooth white chocolate. Top a still-warm serving with a nice scoop of vanilla ice cream for an even bigger treat!

MAKES ONE 12-INCH COOKIE (8 SERVINGS)

14 tablespoons (1¾ sticks/200 grams) butter, softened

1¼ cups (213 grams) dark brown sugar

3 tablespoons granulated sugar

2 large eggs, at room temperature

2 teaspoons pure vanilla extract

2½ cups (350 grams) all-purpose flour

¾ teaspoon baking soda

½ teaspoon salt

1 cup (175 grams) pecans, toasted and chopped

1 cup (175 grams) roughly chopped white chocolate

Vanilla ice cream, for serving

1 Preheat the oven to 350°F (180°C). Butter a 12-inch cast-iron skillet.

2 In a large bowl, cream the butter, brown sugar, and granulated sugar together until light and fluffy. You can do this by hand or with an electric mixer.

3 Add the eggs and vanilla and beat until thoroughly combined.

4 In a separate bowl, whisk together the flour, baking soda, and salt, then gradually mix the dry ingredients into the butter mixture.

5 Fold in the pecans and white chocolate chunks by hand, reserving a few of each to be sprinkled on top.

6 Spread the dough evenly into your prepared skillet and top with the reserved chocolate and nuts.

7 Bake for 18 to 22 minutes, until the edges are slightly crisp but the cookie is still slightly soft and underdone in the middle.

8 Let cool for 30 minutes, then serve while still warm, with scoops of vanilla ice cream. Store leftovers, covered, at room temperature for up to 3 days.

COOL and CREAMY LIME CUSTARD PIE

I love baking with lemons, oranges, and grapefruit, but limes are hands down my most beloved citrus. Limes are a fantastic tart, acidic fruit that gives any dessert a zesty, tropical flavor. To avoid any mouth-puckering bites, I was careful to make sure the sweetness of this pie balances the sour, so you get the best of both worlds in every bite!

**MAKES ONE 9-INCH PIE
(8 SERVINGS)**

FOR THE CRUST

2½ cups (252 grams) crushed gingersnaps

½ cup (1 stick/115 grams) butter, melted

FOR THE FILLING

1 (14-ounce/397-gram) can sweetened condensed milk

¾ cup (180 milliliters) fresh lime juice

3 large egg yolks, at room temperature

¼ cup (57 grams) full-fat sour cream

1 recipe Whipped Cream (page 232), for serving

Lime slices, for serving

1 Preheat the oven to 350°F (180°C).

2 **Make the crust:** Put the gingersnaps in a medium bowl and pour the melted butter over them a little at a time, mixing just until the mixture resembles wet sand. (Don't add all the butter at once, as your cookies might not need all the butter to form a crust.)

3 Press the cookie crust firmly over the bottom and up the sides of a 9-inch pie dish. Place in the fridge to set.

4 **Make the filling:** In a large bowl, whisk together the condensed milk, lime juice, egg yolks, and sour cream until smooth.

5 Pour the filling into the prepared crust.

6 Bake for 15 to 20 minutes, until the edges are set but the center still jiggles slightly. Let cool on a wire rack for 1 hour, then refrigerate for at least 4 hours, or until set and thoroughly chilled, before serving.

7 Just before serving, top the pie with the whipped cream and decorate with lime slices. Serve cold. Store leftovers, covered, in the fridge for up to 3 days.

CINNAMON SEMIFREDDO *with* HONEY TOFFEE SWIRL

A trick of the trade (or at least, a trick from someone who isn't opposed to indulging her sweet tooth): Always have a frozen dessert on hand for any time you *need* something sweet but don't have the *time* to make something sweet.

The combination of honey, cinnamon, and toffee makes this dessert sing. It's a flavor combination that never goes wrong, and when it's in a semifreddo? *Perfection.*

MAKES 10 SERVINGS

- 2 large eggs, at room temperature
- 8 large egg yolks, at room temperature
- 1 cup (170 grams) dark brown sugar
- 1 tablespoon pure vanilla extract
- 2⅓ cups (551 milliliters) heavy cream
- ½ cup (142 grams) Honey Toffee Sauce (page 230), plus more for topping, if desired
- 1 teaspoon ground cinnamon
- 1 cup (142 grams) walnuts, toasted and chopped

1 Line a 9 x 5-inch loaf pan with three pieces of plastic wrap, leaving a few inches hanging over the edges on all sides.

2 In a medium saucepan, bring 2 inches of water to a simmer over low heat. Set a strainer over a bowl and set that next to you at the stove.

3 In a separate heat-safe bowl, whisk together the eggs, egg yolks, brown sugar, and vanilla. Place the bowl over the saucepan of simmering water and cook, whisking continuously, until the mixture is thickened and the sugar has dissolved. Immediately pour the mixture through the strainer into the bowl and let cool to room temperature.

4 In a separate large bowl using a handheld mixer, whip the cream to soft peaks, then fold in the cooled egg mixture until evenly combined.

5 Drizzle the honey toffee sauce on top of the cream-egg mixture, then sprinkle on the cinnamon and nuts. Fold just once or twice to create a ripple (you don't want to mix this evenly).

6 Pour the mixture into the prepared loaf pan. Cover with the overhanging plastic wrap and freeze for at least 6 hours or up to overnight.

7 When ready to serve, unwrap the semifreddo and invert it onto a serving plate. Slice and serve, topped with a touch more honey toffee sauce, if desired.

STRAWBERRY DUMP CAKE

When it comes to whipping up a dessert on weeknights, simplicity is key—and it doesn't get much simpler than this dump cake! It's quick, easy, and delicious. All you have to do is scatter your ingredients into your baking dish and watch it cook up golden, crisp, and *so* delicious!

MAKES ONE 9-INCH CAKE (8 SERVINGS)

5 cups (25 ounces/ 710 grams) fresh strawberries, halved

1 cup (225 grams) plus 2 tablespoons granulated sugar

1 tablespoon cornstarch

1¼ cups (177 grams) all-purpose flour

1 teaspoon baking powder

¼ teaspoon salt

¾ cup (1½ sticks/170 grams) butter, melted

Vanilla ice cream, for serving

1 Preheat the oven to 375°F (190°C).

2 In a large bowl, combine the strawberries, 2 tablespoons of the sugar, and the cornstarch and pour into a 9-inch round baking dish.

3 In a small bowl, whisk together the flour, remaining 1 cup (225 grams) sugar, the baking powder, and the salt and sprinkle this mixture over the strawberries (do not stir).

4 Drizzle the melted butter over the flour mixture (again, without stirring).

5 Bake for about 30 minutes, then check to see if there are any visible dry bits of flour; if so, push them down into the strawberries with the back of a spoon. Bake for another 25 to 30 minutes, until golden. Let cool for 20 minutes before serving.

6 Serve warm, with a scoop of vanilla ice cream. Store leftovers, covered, in the fridge for up to 2 days.

PECAN PIE COBBLER

My pecan pie cobbler tastes of the holidays in every bite, but I bake this all year round and make no apologies! This cobbler is packed with wonderful caramel flavors and sweet, crunchy, toasted pecans—it has all the tastes of a perfectly baked Thanksgiving pecan pie but without all the added fuss!

MAKES ONE 9-INCH COBBLER (8 SERVINGS)

½ cup (1 stick/115 grams) butter, melted

1¼ cups (177 grams) all-purpose flour

¾ cup (170 grams) granulated sugar

1 tablespoon baking powder

¼ teaspoon salt

⅔ cup (142 milliliters) whole milk

1 teaspoon pure vanilla extract

1½ cups (213 grams) pecans, toasted and coarsely chopped

1⅓ cups (227 grams) dark brown sugar

1½ cups (360 milliliters) boiling water

Vanilla ice cream, for serving

1 Preheat the oven to 350°F (180°C). Pour the melted butter into a 9-inch round baking dish.

2 In a large bowl, combine the flour, granulated sugar, baking powder, and salt, then stir in the milk and vanilla until just combined.

3 Pour the batter into the baking dish on top of the melted butter, then evenly sprinkle with the pecans and brown sugar. Pour the boiling water over the whole thing and leave it without stirring.

4 Bake for 35 to 40 minutes, until golden brown. Let cool for 45 minutes (it will thicken slightly).

5 Serve warm, with a big scoop of vanilla ice cream. Store leftovers, covered, at room temperature for up to 2 days.

GOOEY JAM TART

Like a mother loves her children, I love all the desserts I create equally—but if someone were to twist my arm, I'd have to admit that this gooey jam tart is hands down an all-time favorite!

This tart has only a few ingredients and can be made in a small amount of time. With their buttery, crispy pastry base and soft, sweet filling, jam tarts are a master class in how different flavors and textures can turn a humble dessert into a favorite.

MAKES ONE 9-INCH TART (8 SERVINGS)

- 1 recipe Pie Crust (page 222)
- ¾ cup (1½ sticks/170 grams) butter, softened
- ½ cup (115 grams) granulated sugar
- ½ cup (90 grams) dark brown sugar
- 2 teaspoons pure vanilla extract
- ½ teaspoon salt
- 2 large eggs, at room temperature
- ½ cup (71 grams) all-purpose flour
- ⅔ cup (225 grams) raspberry jam
  Vanilla ice cream, for serving

1 Preheat the oven to 325°F (165°C).

2 On a floured surface, roll out the pie dough into a round about ⅛ inch thick. Transfer the dough to a 9-inch pie pan. Refrigerate while you make the filling.

3 In a large bowl, cream together the butter, granulated sugar, brown sugar, vanilla, and salt until combined. Mix in the eggs one at a time, then add the flour and stir until just combined.

4 Spread the jam evenly over the bottom of the pie crust. Carefully pour the batter on top of the jam, spreading it evenly.

5 Bake for 50 to 60 minutes, until the center is just set.

6 Place the pie on a wire rack to cool for at least 4 hours before slicing and serving. Don't slice early, or you will have a gooey mess.

7 Serve warm, with a big scoop of vanilla ice cream. Store leftovers, covered with plastic wrap, at room temperature for up to 3 days.

Chapter 4

DINNER PARTY DESSERTS

When my siblings and I were little, we were told to stay out of the dining room where the grown-ups were eating, but we always found a way to sneak in and salvage any fancy dessert we could find. Now that I'm the adult, I get to make those desserts any time I want—and you'll find some of my favorites right here!

To me, a dinner party dessert needs to be elegant, with an air of sophistication (even when it is actually easy to create). All the recipes you'll find in this chapter have indulgent, decadent flavors and look beautiful when placed on the table.

WHITE CHOCOLATE *and* PASSION FRUIT CHEESECAKE

As a pastry chef, I'm not ashamed to say that I *love* white chocolate! I have ever since I was young. It's sweet, creamy, and pairs so well with so many flavors—including tropical passion fruit. If you haven't had this incredible combination before, hang on to your hat, because you are about to take a trip to flavor town!

MAKES ONE 9-INCH CHEESECAKE (12 SERVINGS)

FOR THE CRUST

- 2½ cups (252 grams) crushed graham crackers
- ½ cup (1 stick/115 grams) butter, melted

FOR THE FILLING

- 2¼ cups (597 grams) cream cheese, softened
- 1 cup (225 grams) sour cream, at room temperature
- ¾ cup (170 grams) granulated sugar
- 1 tablespoon fresh lemon juice
- 3 large eggs, at room temperature
- 1½ cups (9 ounces/ 255 grams) chopped white chocolate, melted and cooled
- ½ cup (120 milliliters) strained passion fruit pulp (from about 6 passion fruits)
  Whipped Cream (page 232), for serving

1 Preheat the oven to 325°F (165°C).

2 **Make the crust:** In a medium bowl, combine the crushed graham crackers with the melted butter. Press the mixture firmly over the bottom of a 9-inch springform pan. Set aside.

3 **Make the filling:** In the bowl of a stand mixer fitted with the paddle attachment or in a large bowl using a handheld mixer, beat the cream cheese and sour cream until thoroughly combined.

4 Add the sugar and lemon juice, then beat in the eggs one at a time until blended. Add the melted white chocolate and passion fruit pulp and beat until incorporated.

5 Pour the filling into the pan over the graham cracker crust.

6 Bake for about 1 hour, until firm but still slightly jiggly in the center.

7 While the cheesecake is still warm, run a thin knife around the edge to loosen it from the side of the pan (but don't remove it from the pan). This helps prevent cracks. Let it cool to room temperature in the pan, then refrigerate until fully chilled, at least 4 hours or up to overnight.

8 When ready to serve, remove the springform ring and transfer the cake to a serving dish. Serve with whipped cream. Store leftovers, covered, in the fridge for up to 3 days.

YOGURT COEUR À LA CRÈME *with* MACERATED CHERRIES

This unique French dessert is no fuss to prepare; really, all the work is in the hanging. I pair the tangy cream with in-season macerated dark cherries to give it a sweet flavor and beautiful contrast of color. Don't be afraid to experiment with other seasonal fruits in this lovely dish.

MAKES 8 SERVINGS

FOR THE COEUR À LA CRÈME

- 2 cups (450 grams) full-fat plain Greek yogurt
- ½ cup (115 grams) mascarpone cheese
- ½ cup (120 milliliters) heavy cream
- ⅓ cup (71 grams) granulated sugar
- 1 tablespoon fresh lemon juice

FOR THE MACERATED CHERRIES

- 3 cups (15 ounces/ 426 grams) fresh cherries, halved and pitted
- 3 tablespoons granulated sugar
- 1 tablespoon grated orange zest
- 2 tablespoons fresh orange juice
- 1 tablespoon fresh lemon juice
- 1 tablespoon brandy (optional)

1 **Make the coeur à la crème:** Place a strainer over a large bowl. Line the strainer with four layers of cheesecloth.

2 In a medium bowl, stir together the yogurt, mascarpone, cream, sugar, and lemon juice until combined.

3 Pour the mixture into the prepared strainer, then gather up the corners of the cheesecloth and tie them in a knot.

4 Place the yogurt mixture with the cheesecloth, strainer, and the bowl in the refrigerator to drain for at least 6 hours or up to overnight.

5 **Make the macerated cherries:** About 30 minutes before you are ready to serve, in a medium bowl, combine the cherries, sugar, orange zest, orange juice, lemon juice, and brandy (if using). Let the cherries sit, stirring occasionally, until the sugar has dissolved and the cherries have released some juice, roughly 30 minutes.

6 Remove the coeur à la crème from the cheesecloth and place it in a serving dish. Pour the macerated cherries around the coeur à la crème and serve immediately. Store leftovers, covered, in the refrigerator for up to 2 days.

DECADENT COCOA PANNA COTTA

Panna cotta, a traditional Italian dessert, is something I could happily eat after dinner (or before, I don't judge, and I assume you won't either!) for the rest of my days. I love adding rich bittersweet chocolate to panna cotta; it tastes like the silkiest, smoothest chocolate mousse you could ever imagine.

MAKES 6 SERVINGS

¼ cup (60 milliliters) hot water

4 teaspoons unflavored powdered gelatin

2 cups (480 milliliters) heavy cream

1 cup (240 milliliters) whole milk

¾ cup (170 grams) granulated sugar

1 cup (115 grams) unsweetened cocoa powder, sifted

2 teaspoons pure vanilla extract

Whipped Cream (page 232), for serving

1 Set six 4-ounce serving dishes on a small tray.

2 Place the hot water in a small bowl and sprinkle on the gelatin without stirring. Set aside for a few minutes to bloom, until the gelatin is spongy.

3 In a medium saucepan, combine the cream, milk, sugar, and cocoa powder and bring to a simmer over medium heat, stirring frequently. Remove from the heat, add the gelatin mixture and vanilla, and whisk to combine.

4 Strain the cocoa mixture into a large bowl to remove any lumps and ensure it is silky smooth.

5 Evenly divide the mixture among the serving dishes. Refrigerate for at least 6 hours, or until chilled and set.

6 Serve topped with a little whipped cream. Store leftovers, covered, in the fridge for up to 3 days.

ELEGANT TIRAMISU CREPE CAKE

If you're looking for a showstopping dessert for a dinner party, this tiramisu crepe cake will make your guests' jaws drop! This gorgeous crepe cake has sixteen beautifully thin layers with a generous filling of mascarpone cream in between each one. With every bite, you get a little hit of booze, cream, and coffee—all the ingredients you need for a successful dinner party!

MAKES 10 SERVINGS

FOR THE ESPRESSO CREPES

- 1   recipe Crepes (page 225)
- 1   tablespoon instant espresso powder

FOR THE COFFEE SYRUP

- ½   cup (120 milliliters) warm water
- ¼   cup (60 milliliters) rum
- 3   tablespoons granulated sugar
- 1   tablespoon instant espresso powder
- 1   tablespoon unsweetened cocoa powder

FOR THE FILLING

- 2   cups (675 grams) mascarpone cheese, at room temperature
- 1   cup (225 grams) granulated sugar
- 2   tablespoons rum (optional)
- 3   cups (720 milliliters) heavy cream

TO ASSEMBLE

- ⅓   cup (71 milliliters) heavy cream
-     Cocoa powder, for dusting

1   **Make the espresso crepes:** Prepare the crepe batter as directed on page 225, then add the instant espresso and stir until incorporated. Chill and cook the batter as directed. You should have 16 espresso crepes; let them cool completely before using.

2   **Make the coffee syrup:** In a 9-inch baking dish, combine the warm water, rum, sugar, instant espresso, and cocoa powder and stir until the dry ingredients have dissolved. Set aside.

3   **Make the filling:** In a medium bowl using a handheld mixer, whip the mascarpone, sugar, and rum (if using) to combine. Add the heavy cream and whip until medium-soft peaks form.

4   **Assemble the cake:** Briefly dip 1 crepe in the syrup and place on a serving plate. Spread about ⅓ cup (2½ ounces/71 grams) of the mascarpone filling over the crepe. Dip a second crepe in the syrup, place it on top of the first crepe, and spread with ⅓ cup (2½ ounces/71 grams) of the filling. Repeat until you have used all the crepes; do not top the final crepe with mascarpone filling.

5   Brush the top crepe with any remaining syrup. Cover and refrigerate for at least 6 hours or up to overnight to allow the crepes to absorb the syrup and the flavors to develop.

6   When ready to serve, in a medium bowl using a handheld mixer, whip the heavy cream to soft peaks. Dollop the whipped cream on top of the cake and sprinkle with a generous dusting of cocoa powder. Slice and serve! Store leftovers, covered, in the fridge for up to 3 days.

CHÈVRE and HONEY TART with BLUEBERRY COMPOTE

Chèvre just isn't a savory cheese to include on your cheeseboard. When baked, this cheese becomes so creamy and takes on that lovely, sharp taste we love from a good goat cheese. Kick your dinner party up a notch with this sweet and subtle tart—I bet you'll get people talking!

MAKES ONE 9-INCH TART (8 SERVINGS)

- 1 recipe Pie Crust (page 222)
- ⅔ cup (142 grams) cream cheese, softened
- ⅔ cup (142 grams) sour cream, at room temperature
- ½ cup (115 grams) chèvre (fresh goat cheese), softened
- 3 tablespoons honey
- 2 tablespoons granulated sugar
- 2 large eggs, at room temperature
- 1 teaspoon grated lemon zest
- 1 teaspoon fresh lemon juice
- ½ teaspoon pure vanilla extract
- ¼ teaspoon salt
- 1 recipe Blueberry and Lemon Compote (page 234), for serving

1 Preheat the oven to 375°F (190°C).

2 On a floured surface, roll out the pie dough into a round about ⅛ inch thick. Transfer the dough to a fluted 9-inch tart pan with a removable bottom, then line it with parchment paper and fill with pie weights (or dried beans).

3 Blind-bake the crust for 15 minutes, then remove the parchment and pie weights and bake for another 10 minutes, until the bottom of the crust is baked through. Remove from the oven and set aside. Reduce the oven temperature to 325°F (165°C).

4 In a large bowl using a handheld mixer, beat the cream cheese, sour cream, chèvre, honey, and sugar together until smooth. Add the eggs one at a time and beat until evenly mixed. Add the lemon zest, lemon juice, vanilla, and salt and mix until combined.

5 Pour the filling into the prebaked tart shell.

6 Bake for about 20 minutes, until the edges are set but the center is still a bit jiggly. Let cool for 1 hour at room temperature, then refrigerate until completely cooled, about 3 hours more.

7 To serve, slice and pour some blueberry and lemon compote over the top. Store leftovers, covered, in the fridge for up to 2 days.

BOOZY CHOCOLATE *and* PRUNE CAKE

When I was young, boozy desserts were for adults only. Now that I am an adult (occasionally, if the situation calls for it!), I try to incorporate liquor into my desserts when I can. This decadent cake is not only sophisticated but also has a nice kick at the end, which I love.

MAKES ONE 9-INCH CAKE (8 SERVINGS)

1½ cups (7½ ounces/ 213 grams) pitted prunes, finely chopped

½ cup (120 milliliters) dark rum

2⅓ cups (14 ounces/ 397 grams) chopped bittersweet chocolate

1¼ cups (2½ sticks/ 284 grams) butter, diced, softened

4 large eggs, at room temperature

6 large egg yolks, at room temperature

½ cup (115 grams) granulated sugar

¼ teaspoon ground cinnamon

¼ teaspoon ground cardamom

1½ cups (213 grams) all-purpose flour

1 recipe Chocolate-Butter Glaze (page 237), pouring consistency

Vanilla ice cream, for serving

1 Preheat the oven to 350°F (180°C). Butter a 9-inch round cake pan and line it with parchment paper.

2 In a small saucepan or in a microwave-safe bowl, combine the prunes and rum and heat until the rum is steaming. Cover to prevent evaporation and set aside to cool and let the prunes plump up.

3 In another small saucepan or in a microwave-safe bowl, melt the chocolate and butter together. Allow to cool.

4 In a separate bowl, beat the eggs, egg yolks, sugar, cinnamon, and cardamom together until creamy.

5 Beat the cooled rum and chocolate mixtures into the eggs, then fold in the flour until just fully combined. Spread the batter evenly in the prepared cake pan.

6 Bake for 30 to 35 minutes, until a skewer inserted into the center comes out with some moist crumbs attached. The cake should seem slightly underdone.

7 Let cool for about 30 minutes before drizzling with the glaze. Serve warm, with vanilla ice cream. Store leftovers, covered, at room temperature for up to 3 days.

SPARKLING ROSÉ *and* RASPBERRY GRANITA

When we were testing this recipe, my culinary assistant, Ami, asked me to sacrifice a bottle of my sparkling rosé. I have to admit, I was reluctant to oblige. You think I would have learned not to doubt her by now—Ami has yet to steer me wrong! This granita is such a special sweet treat for the grown-ups to enjoy after dinner.

MAKES 8 SERVINGS

2½ cups (12½ ounces/ 355 grams) raspberries, fresh or frozen

½ cup (115 grams) granulated sugar

1 (750-milliliter) bottle fruity sparkling rosé wine

1 In a medium saucepan, combine the raspberries and sugar and heat over low heat, stirring frequently, until the raspberries are soft and juicy and the sugar has dissolved.

2 Strain the raspberry mixture through a fine-mesh strainer into a 9 x 13-inch metal baking pan, pressing on the solids with a spatula to extract as much juice as you can. Discard the solids.

3 Pour the rosé into the pan and stir to combine with the raspberry juice.

4 Place the pan in the freezer. After an hour, use a fork to scrape the frozen crystals from the bottom and edges of the pan into the center. Do this every hour for 4 to 5 hours total, until the granita is fully frozen into icy shards.

5 Serve immediately or store in an airtight container in the freezer for up to 7 days.

DULCE DE LECHE LAVA CAKE

I have yet to meet a person who would decline a molten lava cake (I'm sure they exist, but I've never met a leprechaun, either!). Instead of the traditional molten dark chocolate, I use sweet dulce de leche for these cakes. It's a wonderful twist on a classic favorite and one that will really wow your guests.

2   large eggs, at room temperature

2   large egg yolks, at room temperature

1   teaspoon pure vanilla extract

¼   teaspoon salt

1   (14-ounce/397-gram) can dulce de leche

¼   cup (35 grams) all-purpose flour

Vanilla ice cream, for serving

1   Preheat the oven to 425°F (220°C). Generously butter four 6-ounce ramekins and place them on a baking sheet.

2   In a large bowl using a handheld mixer, whisk together the eggs, egg yolks, vanilla, and salt for a few minutes, until thick and pale yellow.

3   Whisk in the dulce de leche until smooth, then fold in the flour until just combined.

4   Divide the batter evenly among the prepared ramekins.

5   Bake for 20 to 23 minutes, until the cakes are still soft and a little jiggly in the center. (Take care not to overcook the cakes, or they won't have a lava center.) Allow to rest for 2 minutes before serving.

6   Serve hot, with a scoop of vanilla ice cream. Store leftovers, covered, in the fridge for up to 2 days.

SILKY COFFEE PUDDING

This is a humble re-creation of one of my favorite desserts from a very special restaurant in Los Angeles called Cassia. If you love a good morning cappuccino, this one is for you. Everything about this pudding speaks to me: it's bitter, sweet, creamy, and unbelievably silky smooth. Serve it in individual dishes for a simple, elegant presentation!

MAKES 8 SERVINGS

- 2 cups (480 milliliters) whole milk
- ½ cup (120 milliliters) strong black coffee
- ½ cup (120 milliliters) heavy cream
- 1 large egg, at room temperature
- 2 large egg yolks, at room temperature
- ½ cup (85 grams) dark brown sugar
- ¼ cup (28 grams) cornstarch
- 2 tablespoons (28 grams) butter, softened
- 1 teaspoon instant espresso powder
- ½ teaspoon salt
- 1 recipe Whipped Cream (page 232), for serving

1 Set a sieve over a bowl and set it next to the stove. Place eight 6-ounce serving dishes on a tray.

2 In a medium saucepan, combine the milk, coffee, and cream. Slowly bring to a simmer over medium heat.

3 Meanwhile, in a medium bowl, whisk together the egg, egg yolks, brown sugar, and cornstarch until blended.

4 While whisking continuously, ladle some of the hot milk mixture over the eggs and whisk to combine (this is called tempering, and it prevents the hot liquid from scrambling the eggs). Still whisking, slowly pour in the rest of the hot milk, then return the custard mixture to the saucepan.

5 Over medium-low heat, whisk the pudding continuously for 6 to 8 minutes, until it has thickened enough to coat the back of a spoon. Immediately remove from the heat and strain it into the bowl to stop the cooking and remove any lumps.

6 Whisk in the butter, instant espresso, and salt. Taste and, if needed, add a little more instant espresso for an even stronger coffee flavor.

7 Divide the pudding evenly among the serving dishes, cover, and refrigerate for several hours, until cold and set.

8 Serve topped with whipped cream. Store leftovers, covered, in the fridge for up to 3 days.

COCONUT CRÈME BRÛLÉE

Crème brûlée is a relatively simple dessert, but there is nothing more satisfying than cracking through that crispy caramelized sugar topping into the smooth, rich custard below. Or that's what I thought, until I added coconut to the mix. The subtle, unexpected tropical flavors transform this dessert into a decadent treat that instantly transports me to the idyllic island of my dreams!

MAKES 6 SERVINGS

1 (14-ounce/403-milliliter) can full-fat coconut cream

1 cup (240 milliliters) heavy cream

½ cup (115 grams) plus 6 tablespoons (43 grams) granulated sugar

1 teaspoon grated lime zest

1 tablespoon fresh lime juice

6 large egg yolks, at room temperature

1 Preheat the oven to 325°F (165°C). Place six 4-ounce ramekins in a 9 x 13-inch roasting pan. Bring a kettle of water to a boil.

2 Open the can of coconut cream and scoop the opaque white coconut solids into a medium saucepan; discard the clear liquid remaining in the can. Add the heavy cream and ½ cup (115 grams) of the sugar to the pan and heat over medium-low heat until the sugar has dissolved and the liquid is simmering.

3 Remove from the heat and add the lime zest and lime juice. Put a lid on the pot and let the zest infuse the liquid for 15 minutes.

4 Whisk the egg yolks in a medium bowl, and then, while whisking continuously, slowly pour in ½ cup of the hot coconut milk mixture and whisk to combine (this is called tempering, and it prevents the hot liquid from scrambling the eggs). Still whisking, slowly pour in the rest of the hot coconut milk, then strain the custard mixture into a measuring cup with a pour spout and discard any solids.

5 Divide the custard evenly among the prepared ramekins. Slide the ramekins in the roasting pan into the oven, then carefully pour boiling water from the kettle into the roasting pan so it comes halfway up the sides of the ramekins.

6 Bake for 40 to 45 minutes, until the custard looks set but still has a slight jiggle in the middle. (It may still seem loose but will firm as it cools.) Let cool to room temperature, then cover the custards and refrigerate for at least 4 hours to fully chill. (The chilled custard can be refrigerated for up to 3 days before serving.)

7 When ready to serve, top each custard with 1 tablespoon of the remaining sugar and use a kitchen torch to melt and caramelize the sugar. (Alternatively, you can melt the sugar under a hot broiler for 1 to 2 minutes or until caramelized.) Serve immediately.

GROWN-UPS' RUM RAISIN SEMIFREDDO

This recipe is missing a few steps at the end, so I thought we should address them up here. The final steps are: Send the kids to bed, stand in front of the freezer, and unapologetically eat this semifreddo straight from the dish! Go on. You've been adulting all week. You deserve this!

1 cup (142 grams) raisins

½ cup (120 milliliters) dark rum

2 large eggs, at room temperature

8 large egg yolks, at room temperature

1 cup (170 grams) dark brown sugar

1 tablespoon pure vanilla extract

2⅓ cups (551 milliliters) heavy cream

1 Line a 9 x 5-inch loaf pan with three sheets of plastic wrap, leaving a few inches hanging over the edges on all sides.

2 In a small saucepan, combine the raisins and rum and bring to a simmer over low heat. Remove the saucepan from the heat and cover with a lid. Let stand for 20 minutes to allow the raisins to soak up the rum.

3 In a double boiler, combine the eggs, egg yolks, brown sugar, and vanilla and cook over low heat, whisking continuously, until the mixture has thickened and the sugar has dissolved. Immediately pour the mixture through a fine-mesh sieve into a bowl and let cool to room temperature.

4 In the bowl of a stand mixer fitted with the whisk attachment or in a medium bowl using a handheld mixer, whip the cream to soft peaks. Fold in the cooled egg mixture and the rum and raisins until evenly combined.

5 Spoon the mixture into your prepared loaf pan and spread it evenly. Cover with the overhanging plastic wrap and freeze for at least 6 hours or up to overnight.

6 When ready to serve, unwrap the semifreddo and invert it onto a serving plate, then slice and serve. Store leftovers in an airtight container in the freezer for up to 4 weeks.

DEATH BY CHOCOLATE CAKE

This cake has serious wow factor: you'll leave your guests speechless when you place this masterpiece of a dessert in the center of the table. But don't be intimidated by the somewhat morbid name—it just means that this incredibly chocolaty cake tastes so good, you'll think you've died and gone to heaven!

MAKES ONE 9-INCH LAYER CAKE (10 SERVINGS)

FOR THE WHIPPED DARK CHOCOLATE GANACHE

- 2 cups (12 ounces/340 grams) chopped bittersweet chocolate
- 2 cups (480 milliliters) heavy cream
- 2 tablespoons (28 grams) butter, softened

FOR THE CAKE

- ½ cup (57 grams) black cocoa powder (see note)
- ⅓ cup (2 ounces/57 grams) very finely chopped bittersweet chocolate
- 1 cup (240 milliliters) hot strong-brewed coffee
- 2 cups (340 grams) dark brown sugar
- ½ cup (1 stick/115 grams) butter, softened
- ½ cup (120 milliliters) vegetable oil
- ½ cup (115 grams) sour cream, at room temperature
- 3 large eggs, at room temperature
- 2 teaspoons pure vanilla extract
- 1¾ cups (247 grams) all-purpose flour
- ¼ cup (28 grams) cornstarch
- ½ teaspoon salt
- 1 teaspoon baking soda

1 **Make the whipped dark chocolate ganache:** Place the chocolate in a heatproof bowl.

2 In a small saucepan, warm the cream over medium heat until just simmering, then pour it over the chocolate. Let stand, without stirring, for 10 minutes, then use a whisk to gently stir the cream and chocolate together until you have a smooth, even ganache.

3 Stir in the butter until melted and combined, then set the ganache aside to cool completely and thicken while you make the cake.

4 **Make the cake:** Preheat the oven to 350°F (180°C). Butter two 9-inch round cake pans and line them with parchment paper.

5 In the bowl of a stand mixer or in a large bowl, combine the cocoa powder, chocolate, and hot coffee. Stir, then cover and let stand for 10 minutes to melt the chocolate.

6 Add the brown sugar, butter, oil, sour cream, eggs, and vanilla. Using the stand mixer's paddle attachment or a handheld mixer, whip on medium-high speed for several minutes until thoroughly mixed and aerated.

7 In a medium bowl, whisk together the flour, cornstarch, salt, and baking soda, then add the dry ingredients to the chocolate mixture in two or three increments, scraping down the sides and bottom of the bowl after each addition. Stop just as soon as the batter is evenly mixed.

8 Divide the batter evenly into the prepared pans.

9 Bake for 35 to 40 minutes, until a toothpick inserted into the center comes out clean. Let the cakes cool in the pans for 10 minutes, then invert them onto a wire rack to cool completely.

CONTINUES

DEATH BY CHOCOLATE CAKE

10 **Assemble the cake:** Once the ganache has cooled and thickened to the consistency of thick custard, it is ready to whip. In the bowl of a stand mixer fitted with the whisk attachment or using a handheld mixer, whip the ganache on medium-high speed for just a few minutes, until fluffy. (Take care not to overwhip or your ganache can become grainy.)

11 Cut the cooled cakes in half horizontally to create 4 layers total. Place one layer on a serving plate and frost generously with ganache. Repeat with the remaining layers, then frost the top and sides of the cake.

12 With an offset spatula gently scrape the ganache off the sides, leaving a thin layer of frosting with some cake showing through.

13 Slice and enjoy! Store leftovers, covered, at room temperature for up to 3 days.

Note: Black cocoa is a deep, heavily dark Dutch-process cocoa powder. If you don't have access to it, you can use regular unsweetened Dutch-process cocoa powder and still get amazing results.

LAYERED PAVLOVA with SUMMER FRUIT and ROSE

I pull out the big guns when it comes to entertaining. When I bring my dessert to the table, it has to have that WOW factor! This layered pavlova is piled high with fruit and cream, and there's a touch of rose water in the whipped cream that takes all those flavors to another level. Once your guests have a bite of this masterpiece, they'll be begging for the recipe.

MAKES 8 SERVINGS

1 recipe Basic Meringue (page 220)

FOR THE ROSE WATER WHIPPED CREAM

3 cups (720 milliliters) heavy cream

3 tablespoons rose water

2 tablespoons honey

TO ASSEMBLE

4 peaches (20 ounces/ 568 grams), pitted and sliced

1 recipe Strawberry Compote (page 234)

1 Preheat the oven to 225°F (110°C). Line two baking sheets with parchment paper.

2 Spoon the meringue into three mounds total on the prepared baking sheets, dividing it evenly, and spread each mound with a spatula into a 9-inch round.

3 Bake for 80 to 90 minutes, until the meringue is dry to the touch, then turn the oven off and let the meringue cool in the oven for 2 hours. (The meringue can be stored, covered with plastic wrap, at room temperature for up to 2 days before serving.)

4 **Make the rose water whipped cream:** Using a stand mixer or electric hand mixer, whip the cream, rose water, and honey on high speed until stiff peaks form. Set aside.

5 Place one meringue on a serving platter and add about one-third of your whipped cream. Lay half the sliced peaches on top and drizzle generously with compote. Place a second meringue on top and layer with half the remaining whipped cream, the remaining peaches, and another drizzle of compote. Place the final meringue on top, spread with the remaining whipped cream, and garnish with a little fresh fruit.

6 Serve straight away. Store any leftovers in an airtight container in the fridge for up to 1 day.

UNBELIEVABLE PEAR *and* DARK CHOCOLATE CRISP

Call it divine inspiration, but I felt like I was struck by a bolt of lightning for this dessert! "How good could it be," you ask? I dare to say, this is one of my all-time favorite desserts—and I'm not even a fan of pears. I urge you, do not pass this recipe by!

MAKES ONE 9-INCH CRISP (6 SERVINGS)

FOR THE CHOCOLATE CRISP TOPPING

- 1 cup (142 grams) all-purpose flour
- 2 tablespoons unsweetened cocoa powder
- ¾ cup (128 grams) dark brown sugar
- ¼ teaspoon salt
- ¾ cup (1½ sticks/170 grams) cold butter, diced
- ½ cup (3 ounces/ 85 grams) finely chopped bittersweet chocolate

FOR THE PEAR FILLING

- 7 or 8 large ripe pears (about 2 pounds/ 900 grams), peeled, cored, and chopped into 2-inch chunks
- 1 tablespoon cornstarch
- 2 teaspoons fresh lemon juice

  Vanilla ice cream, for serving

1. Preheat the oven to 350°F (180°C).

2. **Make the chocolate crisp topping:** In a large bowl, lightly whisk together the flour, cocoa powder, brown sugar, and salt.

3. Add the butter and, using your fingertips, rub it into the dry ingredients until the mixture resembles coarse bread crumbs. Mix in the chocolate, then cover and refrigerate until ready to bake your crisp. (The topping can be refrigerated for up to 4 days or even frozen for up to 4 weeks before using.)

4. **Make the pear filling:** In a large bowl, mix the pears, cornstarch, and lemon juice. Pour the filling into a 9-inch baking dish.

5. Top with the chocolate crisp mixture, distributing it evenly.

6. Bake for 45 to 50 minutes, until the pears are soft and the topping is firm to the touch.

7. Serve warm, with vanilla ice cream. Store leftovers, covered, in the fridge for up to 3 days.

Chapter 5

WEEKEND BRUNCH TREATS

There's a reason champagne is popped for brunch: it's a celebration of good food, good company, and, of course, *the fact that it is finally the weekend!*

Forget the long lines at restaurants—heat up your frying pan and rally the willing pancake flippers in your home! I hope that once you try the recipes in this chapter, brunch becomes a regular weekend occurrence in your household.

NO-YEAST CINNAMON ROLLS

Patience is a virtue. Unfortunately, it isn't a virtue bestowed upon me, especially when it comes to waiting for cinnamon rolls. Luckily, these no-yeast cinnamon rolls don't need time to proof! You can just mix, roll, and eat them in no time flat. Less waiting, more eating . . . that's the ideal weekend brunch.

MAKES 12 ROLLS

FOR THE FILLING

- ½ cup (1 stick/115 grams) butter, melted
- 1¼ cups (213 grams) dark brown sugar
- 2½ tablespoons ground cinnamon

FOR THE DOUGH

- 3 cups (426 grams) all-purpose flour
- ⅓ cup (71 grams) granulated sugar
- 1½ level tablespoons baking powder
- ¾ teaspoon salt
- ¾ cup (1½ sticks/170 grams) cold butter, diced
- 1 cup (225 grams) sour cream, at room temperature
- 2 large egg yolks, at room temperature
- 1½ teaspoons pure vanilla extract

FOR THE CREAM CHEESE GLAZE

- ½ cup (4 ounces/115 grams) cream cheese, softened
- 4 tablespoons (½ stick/ 57 grams) butter, softened
- 1 cup (115 grams) confectioners' sugar
- 1 teaspoon pure vanilla extract

1 Preheat the oven to 400°F (200°C). Butter a 9-inch springform pan and line it with parchment paper.

2 **Make the filling:** In a small bowl, stir together the melted butter, brown sugar, and cinnamon until evenly combined. Set aside.

3 **Make the dough:** In a large bowl, whisk together the flour, granulated sugar, baking powder, and salt.

4 Add the butter and, using your fingertips, rub it into the dry ingredients until the mixture resembles coarse bread crumbs.

5 In a measuring cup, combine the sour cream, egg yolks, and vanilla, then add the wet ingredients to the flour mixture and gently stir until the dough comes together into a smooth ball.

6 Turn the dough out onto a floured work surface and roll it into a 16 x 12-inch rectangle.

7 Spoon the cinnamon filling over the dough, spreading evenly and leaving a 1-inch border along one of the long edges.

8 Beginning from the long edge opposite the edge with the exposed border, tightly roll the dough into a log. Using a sharp, serrated knife, cut the log crosswise into 12 slices, 1½ inches thick each. Place the slices cut-side down in the prepared pan, arranging 9 rolls around the edge of the pan and 3 rolls in the center.

9 Bake for 30 to 35 minutes, until the rolls have puffed up, the filling is bubbling, and the buns are lightly golden. Remove from the oven and let cool in the pan for 10 minutes, then transfer the rolls to a wire rack set over a baking sheet to cool slightly.

10 **Make the cream cheese glaze:** In a large bowl, whisk together the cream cheese, butter, confectioners' sugar, and vanilla until smooth.

11 Generously spoon the glaze over top. Enjoy immediately. Store any leftovers in an airtight container at room temperature for up to 2 days.

OVERNIGHT BELGIAN WAFFLES

I'm a planner. I like to know what's next on the menu, many meals ahead! That's why I love this waffle recipe. I mix up the batter the day before and let it proof in the fridge overnight; then all I have to do in the morning is pour that batter on my waffle iron and wait for it to turn into delicious, crispy Belgian-style waffles.

MAKES 4 WAFFLES

- 2 cups (284 grams) all-purpose flour
- 1½ teaspoons instant yeast
- ¾ teaspoon salt
- 2 large eggs, at room temperature
- 1½ cups (360 milliliters) whole milk
- 6 tablespoons (¾ stick/ 85 grams) butter, melted
- 2 tablespoons pure maple syrup, plus more for serving
- 1 teaspoon pure vanilla extract
- Butter, for cooking and serving

1 In a large, lidded container, combine the flour, yeast, and salt.

2 In a medium bowl, whisk together the eggs, milk, melted butter, maple syrup, and vanilla, then stir this into the dry ingredients. Mix until just combined, with a few small lumps remaining.

3 Cover the batter with the lid and refrigerate for at least 12 hours or preferably overnight. The waffle batter will double in size, so make sure your container is large enough to accommodate this.

4 The next day, heat a waffle iron to medium-high. When it's hot, brush the waffle iron plates with butter, then pour about ¾ cup of batter into the center, close the lid, and cook for about 3 minutes, until golden brown. Transfer the waffle to a plate and keep warm under a clean kitchen towel while you repeat with the remaining batter.

5 Serve warm, with maple syrup and a big piece of butter. Store leftovers in an airtight container in the fridge for up to 2 days.

BANANA CINNAMON WAFFLES

After you taste this recipe, you'll move waffles from your "sometimes" breakfast menu to your weekly rotation! These buttermilk waffles loaded with mashed bananas and toasty cinnamon are guaranteed to be an instant family favorite.

MAKES 8 WAFFLES

3 cups (426 grams) all-purpose flour

¼ cup (43 grams) dark brown sugar

1½ teaspoons ground cinnamon

1 tablespoon baking powder

1 teaspoon baking soda

1 teaspoon salt

3 large bananas (10½ ounces/ 300 grams), peeled

4 large eggs, at room temperature

⅔ cup (142 milliliters) vegetable oil

2½ cups (600 milliliters) buttermilk

2 teaspoons pure vanilla extract

Butter, for cooking and serving

Pure maple syrup, warmed, for serving

1 In a large bowl, combine the flour, sugar, cinnamon, baking powder, baking soda, and salt.

2 In a separate large bowl, mash the bananas with a fork until no chunks remain. Whisk in the eggs, oil, buttermilk, and vanilla.

3 Add the wet ingredients to the dry ingredients and whisk together briefly. (A few lumps are okay.)

4 Heat your waffle iron to medium-high. When it's hot, brush the waffle iron plates with butter, then pour about ¾ cup of batter into the center, close the lid, and cook for 3 to 4 minutes, until golden brown. Transfer the waffle to a plate and keep warm under a clean kitchen towel while you repeat with the remaining batter.

5 Serve with butter and warm maple syrup. Store leftovers in an airtight container in the fridge for up to 2 days.

5-INGREDIENT DUTCH BABY PANCAKE

Serving meals family-style doesn't just mean bringing everyone together at the table—it also means fewer dishes! A Dutch baby is shared straight out of the cast-iron pan it's baked in, making it perfect for entertaining a group.

MAKES ONE 10-INCH PANCAKE (4 SERVINGS)

- 3 large eggs, at room temperature
- ½ cup (120 milliliters) whole milk
- ½ cup (71 grams) all-purpose flour
- 3 tablespoons (42 grams) butter, melted
- ¼ teaspoon salt
  Strawberry Compote (page 234), for serving
  Whipped Cream (page 232), for serving

1 Preheat the oven to 425°F (220°C). Place a 10-inch cast-iron skillet on the middle rack to heat up as well.

2 In a large bowl, whisk together the eggs and milk until well blended. Add the flour, 2 tablespoons (28 grams) of the melted butter, and the salt and whisk until smooth.

3 Carefully brush the bottom and sides of the hot skillet with the remaining 1 tablespoon (14 grams) melted butter, then immediately pour the batter into the skillet.

4 Bake, without opening the oven, for about 20 minutes, until golden brown and well risen. The Dutch baby will deflate a bit when you remove it from the oven, but that is normal.

5 Serve immediately, with strawberry compote and whipped cream. This is best enjoyed the day it is made, but any leftovers can be stored, covered, at room temperature for up to 1 day.

CARROT CAKE PANCAKES *with* CREAM CHEESE FROSTING

Georgie, like his mum, is a pancake aficionado. Though he's still young, we are already impressed by his discerning palate, so I know that if he likes a recipe, it's a keeper. These carrot cake pancakes with cream cheese frosting are both baby and grown-up approved.

MAKES 8 PANCAKES

FOR THE CREAM CHEESE FROSTING

- ½ cup (4 ounces/115 grams) cream cheese, at room temperature
- 2 tablespoons granulated sugar
- ¼ cup (57 milliliters) heavy cream
- 1 teaspoon pure vanilla extract

FOR THE PANCAKES

- 1 cup (142 grams) chopped carrots (about 2 medium carrots)
- 1 large egg, at room temperature
- 1¼ cups (300 milliliters) whole milk
- 2 tablespoons vegetable oil
- 1 teaspoon pure vanilla extract
- 1¼ cups (180 grams) all-purpose flour
- ¼ cup (45 grams) dark brown sugar
- 2 teaspoons baking powder
- 1 teaspoon ground cinnamon
- ¼ teaspoon salt
- 2 tablespoons raisins
  Butter, for cooking
  Pure maple syrup, for serving
  Toasted walnuts, for serving

1 **Make the cream cheese frosting:** In a medium bowl using an electric mixer or by hand, whip the cream cheese with the granulated sugar until fluffy and incorporated. Pour in the cream and vanilla and whip until combined and thickened. Refrigerate until serving.

2 **Make the pancakes:** Pulse the carrots in a food processor until finely chopped. Add the egg, milk, oil, and vanilla and process until well blended.

3 In a large bowl, whisk together the flour, brown sugar, baking powder, cinnamon, and salt.

4 Add the carrot mixture to the dry ingredients and stir until just combined, then fold in the raisins.

5 Heat a skillet over medium-low heat. Once hot, brush the skillet lightly with butter, then scoop about ⅓ cup (75 milliliters) of batter per pancake into the skillet. Cook for 3 to 4 minutes, until bubbles form on the surface and the bottom is golden brown, then flip and cook for another 2 minutes, or until the second side is golden brown. Transfer the pancake to a plate and repeat with the remaining batter.

6 Serve stacks of warm pancakes with a big dollop of the cream cheese frosting on top, a drizzle of maple syrup, and a sprinkle of walnuts. Store any leftovers in an airtight container in the fridge for up to 2 days.

KEVIN'S NO-FUSS BANANA OAT PANCAKES

One Saturday morning, pancakes were needed ASAP, but to my family's horror, there was no flour in the house! So these flourless banana oat pancakes were born out of necessity, but their fantastic flavor and texture have made them a staple in our house, where Kevin will often whip up a batch for all of us on the weekend.

MAKES 8 PANCAKES

3 cups (255 grams) rolled oats

2 teaspoons baking powder

¼ teaspoon salt

4 medium bananas (15 ounces/428 grams), peeled

4 large eggs, at room temperature

2 teaspoons pure vanilla extract

6 tablespoons (¾ stick/ 85 grams) butter, melted

Butter, for cooking

Pure maple syrup, for serving

Sliced banana, for serving

1 Using a food processor or an immersion blender, process the oats until they are finely ground, similar to oat flour.

2 In a medium bowl, whisk together the ground oats, baking powder, and salt.

3 In a large bowl, mash the bananas with a fork until no chunks remain. Whisk in the eggs and vanilla. Add the oat mixture and mix until combined. Add the melted butter and give one last stir to incorporate it.

4 Heat a large skillet over medium-low heat. Once hot, brush the skillet lightly with butter, then scoop about ⅓ cup (75 milliliters) of batter for each pancake into the skillet. Cook for 3 to 4 minutes, until the edges are dry and the bottom is golden brown, then flip the pancakes and cook for another 2 minutes, or until the second side is golden brown. Transfer the pancakes to a plate and repeat with the remaining batter.

5 Serve warm, with maple syrup, banana slices, and even a little butter on top. Store leftovers, covered, in the fridge for up to 2 days.

CLASSIC POPOVERS

It's hard to beat a warm popover for a weekend breakfast! These easy-to-make rolls are buttery, light as air, and, when served with softly whipped cream and jam, absolutely irresistible.

Vegetable oil, for greasing

1¾ cups (250 grams) all-purpose flour

½ teaspoon baking powder

1 teaspoon salt

3 large eggs, at room temperature

1½ cups (360 milliliters) whole milk

Raspberry jam, for serving

1 Preheat the oven to 450°F (230°C). Very generously grease all but one well of a 12-cup muffin tin with oil and place it in the oven to heat up.

2 In a large bowl, combine the flour, baking powder, and salt.

3 Crack in the eggs and whisk as you slowly stream in the milk. There may be some lumps, but that's okay, they work themselves out. Transfer the batter to a large measuring cup with a pour spout and set aside.

4 Once the oven is at temperature, very carefully pull the hot muffin tin out and fill the greased wells almost to the top with batter, dividing it evenly, then swiftly slide the tin back into the oven.

5 Without opening the oven, bake for 15 minutes, then reduce the oven temperature to 375°F (190°C) and bake for another 15 minutes, or until the popovers are golden brown. Remove from the oven and turn the popovers out of the pan.

6 Popovers are best enjoyed freshly baked, so serve them hot and fresh from the oven, with raspberry jam. Store leftovers in an airtight container at room temperature for up to 24 hours.

MAKE-AHEAD FRENCH TOAST CASSEROLE

Being a professional chef, I was trained to be prepared (within an inch of my life!) to eliminate the chances of anything going wrong. Naturally, I now gravitate toward recipes like this make-ahead French toast casserole. I'm able to assemble it in advance, and then the only step I have to perform in the morning (besides brewing some coffee) is popping it into the oven. The benefits of this recipe? A warm and comforting casserole, hot from the oven, and a 100 percent stress-free breakfast.

MAKES 8 SERVINGS

- 4 tablespoons (½ stick/ 57 grams) butter, softened
- 10 (1-inch-thick) slices brioche, store-bought or homemade (page 175)
- 6 large eggs, at room temperature
- 2½ cups (600 milliliters) whole milk
- ½ cup (120 milliliters) heavy cream
- ¼ cup (57 grams) granulated sugar
- ½ teaspoon ground cinnamon
- ¼ teaspoon ground nutmeg
- 1 teaspoon pure vanilla extract
- ½ teaspoon salt
- Grated zest of 1 large orange
- Berries and pure maple syrup, for serving

1 Generously butter a 10-inch baking dish.

2 Spread the butter on one side of the bread slices. Cut the bread into large cubes, crust and all, and add them to the baking dish.

3 In a large bowl, whisk together the eggs, milk, cream, sugar, cinnamon, nutmeg, vanilla, salt, and orange zest.

4 Pour the mixture over the bread and press down gently to help the bread absorb the liquid. Cover and refrigerate overnight.

5 The next morning, preheat the oven to 350°F (180°C).

6 Uncover and bake for 50 to 60 minutes, until puffed, golden, and set in the center.

7 Serve warm, with berries and maple syrup. Store any leftovers, covered, in the fridge for up to 2 days.

OLD-FASHIONED JOHNNYCAKES

Johnnycake, hoecake, Shawnee cake, journey cake, spider cornbread—whatever you call them, these cornmeal flatbreads are delicious in their simplicity. Johnnycakes were popular in the Americas well before Europeans arrived and have since become popular worldwide, where there are as many adaptations as there are names. Enjoy this old-fashioned recipe drizzled with maple syrup and smeared with butter!

MAKES 8 TO 10 JOHNNYCAKES

- 1 cup (142 grams) all-purpose flour
- 1 cup (170 grams) fine ground cornmeal
- 2 tablespoons granulated sugar
- 2 teaspoons baking powder
- ½ teaspoon salt
- 2 large eggs, at room temperature
- 1¼ cups (300 milliliters) buttermilk
- 4 tablespoons (½ stick/ 57 grams) butter, melted

  Butter, for cooking and serving

  Pure maple syrup, for serving

1 In a medium bowl, whisk together the flour, cornmeal, sugar, baking powder, and salt.

2 In a separate medium bowl, whisk the eggs, then add the buttermilk and melted butter and whisk until combined. Add the dry ingredients and stir until just combined.

3 In a large skillet or griddle, melt a little butter over medium-low heat. Scoop ⅓ cup (75 milliliters) of batter into the pan for each johnnycake. Cook for 3 to 4 minutes, until bubbles form on the top and the bottom is nicely browned, then flip and cook for another 2 minutes, or until the other side is golden brown. Transfer the johnnycakes to a plate and repeat with the remaining batter.

4 Serve warm, with butter and maple syrup. Store leftovers in an airtight container in the fridge for up to 2 days.

LEMON-BLUEBERRY RICOTTA HOTCAKES

A big stack of lemon-blueberry ricotta hotcakes is what weekends were made for! Treat yourself (and your sweet tooth) to these thick and incredibly moist pancakes. Sharing is optional!

MAKES 10 PANCAKES

2 cups (300 grams) all-purpose flour

¼ cup (115 grams) granulated sugar

1 level tablespoon baking powder

1 teaspoon baking soda

½ teaspoon salt

1½ cups (340 grams) fresh ricotta cheese

4 large eggs, separated

1½ cups (360 milliliters) buttermilk

1 tablespoon grated lemon zest

Butter, for cooking

Blueberry and Lemon Compote (page 234), for serving

1 In a large bowl, whisk together the flour, sugar, baking powder, baking soda, and salt.

2 In a medium bowl, whisk together the ricotta cheese, egg yolks, buttermilk, and lemon zest.

3 Add the wet ingredients to the dry ingredients and mix until just combined.

4 In a clean bowl using a handheld mixer, whip the egg whites to stiff peaks, then gently fold them into the batter.

5 In a large skillet, melt a little butter over medium-low heat. Pour ½ cup (120 milliliters) of batter for each hotcake into the pan and cook for 3 to 4 minutes, until bubbles form on top and the bottom is nicely browned, then flip and cook for another 2 minutes, or until the other side is golden brown. Transfer to a plate and repeat with the remaining batter.

6 Serve immediately, drizzled with compote. Store leftovers in an airtight container in the fridge for up to 2 days.

GEORGIE'S AUSSIE PIKELETS

I learned early that as a new mum, I need to grab peaceful moments when I can. These little Aussie pancakes buy me at least 15 minutes of peace in the morning as I drink my coffee and become human again. I cook off a batch and freeze them so they're always on hand!

MAKES 25 PANCAKES

2 cups (284 grams) all-purpose flour

⅓ cup (71 grams) granulated sugar

2 teaspoons baking powder

¼ teaspoon salt

2 large eggs, at room temperature

1½ cups (360 milliliters) whole milk

4 tablespoons (½ stick/ 57 grams) butter, melted

Butter, for cooking

Raspberry jam, for serving

Whipped Cream (page 232), for serving

1 In a large bowl, stir together the flour, sugar, baking powder, and salt.

2 In a small bowl, whisk together the eggs, milk, and butter until combined.

3 Add the wet ingredients to the dry ingredients and stir until just combined; a few small lumps are okay.

4 In a large skillet or on a griddle, melt a little butter over medium-low heat. Pour 2-tablespoon scoops of batter into the skillet and cook for 2 to 3 minutes, until bubbles form on the top and the bottom is nicely browned. Flip and cook for another minute, or until the other side is golden brown. Transfer the pancakes to a plate and repeat with the remaining batter.

5 Serve warm, with some raspberry vanilla compote and whipped cream.

RASPBERRY and CREAM CHEESE CREPE CASSEROLE

I adore the balance of textures and flavors in this dish: the soft crepes, salty-sweet raspberry and cream cheese filling, and generous drizzle of zesty, lemony syrup on top. This is one of my ultimate comfort foods, and it's perfect for sharing.

MAKES 8 SERVINGS

- 1 recipe Crepes (page 225)

FOR THE FILLING

- 1 cup (8 ounces/225 grams) cream cheese, softened
- ¼ cup (57 grams) granulated sugar
- 2 teaspoons grated lemon zest
- 1 cup (283 grams) Raspberry Vanilla Compote (page 236)

FOR THE MAPLE GLAZE

- 4 tablespoons (½ stick/ 57 grams) butter
- ¼ cup pure maple syrup
- 2 tablespoons fresh lemon juice

1 Make the crepes and let them cool completely.

2 While the crepes are cooling, preheat the oven to 350°F (180°C). Butter a 9-inch round baking dish.

3 **Make the filling:** In a medium bowl, whisk together the cream cheese, sugar, and lemon zest until light and fluffy.

4 Spread 1 tablespoon of the cream cheese filling over a cooled crepe. On top of the cream cheese, spread 1 tablespoon of the compote. Fold the crepe in half and then in half again, then place in the baking dish. Repeat to fill the remaining crepes, overlapping them in the baking dish to create layers.

5 Bake the casserole for 15 to 20 minutes, until the edges are browning and the crepes and filling are warmed through.

6 **Make the maple glaze:** In a small saucepan, combine the butter, maple syrup, and lemon juice and heat over medium-low heat until the butter has melted and the mixture is hot.

7 Remove the casserole from the oven and generously drizzle the top with the maple glaze. Serve immediately. Store leftovers, covered, in the fridge for up to 3 days.

Chapter 6

LEISURELY WEEKEND RECIPES

When I was young, I spent my weekends baking. I was making cakes, buns, breads, and anything else my little baking mind could whip up from morning to night.

Weekends are made for the recipes I've gathered for this chapter. They may be more time-consuming, but to me, there's no better way to spend your free time—not to mention the delicious reward at the end of all your hard work!

NO-KNEAD CINNAMON RAISIN BREAD

Even after all these years of being a professional chef, I'm still amazed that the base ingredients of bread—flour, sugar, yeast, and salt—can yield endless flavors when paired with just a few simple complementary ingredients. This bread is a perfect example of that beautiful transformation: add some cinnamon and raisins to this yeasted dough, and you have an incredibly delicious loaf that's perfect whether sliced fresh or toasted and buttered the next day.

MAKES 1 LOAF

- 3 cups (425 grams) bread flour
- ¾ cup (106 grams) raisins
- 1½ tablespoons granulated sugar
- 2 teaspoons ground cinnamon
- 1½ teaspoons salt
- ½ teaspoon instant yeast
- ¾ cup (180 milliliters) whole milk
- ¾ cup (180 milliliters) water

NOTE: Want to bake it without a Dutch oven? Set your ball of dough on a baking sheet, cover with plastic wrap, and proof for roughly 1 hour. When the proofing is almost finished, preheat the oven to 400°F (200°C). Uncover and bake the bread for 45 to 50 minutes, until a crisp crust has formed and the loaf sounds hollow when tapped.

1 In a large bowl, whisk together the flour, raisins, sugar, cinnamon, salt, and yeast.

2 In a measuring cup, combine the milk and water and heat gently in the microwave or in a pot on the stove until warm to the touch.

3 Pour the wet ingredients into the dry ingredients and stir with a wooden spoon until the dough comes together and forms a ball. (If it seems a little dry, add a splash more liquid.) Cover the bowl with plastic wrap and a kitchen towel and leave it on the counter for at least 12 hours and up to 18 hours, until the dough is sticky and yeasty and has at least doubled in size.

4 Line a Dutch oven with parchment paper. Scrape the dough out of the bowl onto a floured surface and fold it on itself two times, like a book. Roll the dough into a ball, getting it nice and smooth on the bottom, then place it seam-side down in the prepared Dutch oven, cover with the lid, and let proof for 1 hour at room temperature.

5 About 20 minutes before the end of the proofing time, preheat the oven to 400°F (200°C).

6 Score the top of the dough with a sharp knife, then cover the pot and slide it into the oven.

7 Bake for 30 minutes, then remove the lid and bake for another 20 minutes, or until the bread is golden brown and a nice crust has formed.

8 Let the bread cool completely on a wire rack before slicing and serving with butter. Store the loaf in an airtight container at room temperature for up to 3 days. It also freezes really well for up to 4 weeks.

SWEET PUMPERNICKEL BREAD

I have fond memories of vacationing in Florida when my siblings and I were young. Visiting a classic American steak house chain for dinner is one of those beloved memories that continues to stick out! As a family from Ireland, we had never experienced a restaurant with a catchy jingle dedicated to baby back ribs or a recommendation from the waitstaff to order an entire deep-fried onion. But Little Gemma was drawn toward the steak house's incredible pumpernickel bread. Long have I tried to re-create that bread, and finally, I've cracked it!

MAKES 1 LARGE LOAF

2½ cups (355 grams) all-purpose flour

1½ cups (213 grams) whole wheat flour

2 tablespoons unsweetened cocoa powder

1½ teaspoons instant yeast

1½ teaspoons salt

1¼ cups (300 milliliters) water

¼ cup (71 grams) honey

3 tablespoons (60 grams) molasses

2 tablespoons vegetable oil

1 egg, beaten, for egg wash

¼ cup (21 grams) rolled oats, for garnish

1 Line a baking sheet with parchment paper.

2 In the bowl of a stand mixer fitted with the dough hook (or in a large bowl if kneading by hand), combine the all-purpose flour, whole wheat flour, cocoa powder, yeast, and salt.

3 In a measuring cup, combine the water, honey, molasses, and oil, then pour this into the flour mixture. Mix until a dough is formed, then knead the dough on low speed for about 10 minutes (or for 15 minutes by hand). The dough will be smooth and slightly sticky.

4 Transfer the dough to a lightly greased bowl, cover, and let rise until doubled in size, about 1½ hours.

5 Turn the dough out onto a lightly floured surface and roll it into a smooth 12-inch-long log. Place the log on your baking sheet, seam-side down, and carefully cover with plastic wrap, followed by a clean kitchen towel. Let rise at room temperature until doubled in size, about 1 hour.

6 About 20 minutes before the dough finishes proofing, preheat the oven to 350°F (180°C).

7 Generously brush the dough all over with the egg wash and sprinkle with the oats.

8 Bake for 45 to 55 minutes, until the loaf sounds hollow when tapped.

9 Remove from the oven and wrap it in a clean, dry kitchen towel, then let cool completely before slicing. (This gives the bread a soft crust.)

10 Slice and serve with butter. Store the loaf in an airtight container at room temperature for up to 3 days. This bread also freezes wonderfully for up to 4 weeks.

BRAIDED CHOCOLATE BABKA

I judge a good babka by how heavy it is. I want it to be so chockablock with dark chocolate that the bread falls over when sliced—completely laden with rich, bittersweet chocolate! When you cut a good babka, it shouldn't stay upright; all the weight of the chocolate should make it naturally fall onto your plate. Take the time to bake this over the weekend, so you can start the dough on Saturday and bake it on Sunday.

MAKES 1 LOAF

FOR THE DOUGH

- 2¼ cups (319 grams) all-purpose flour
- ¼ cup (57 grams) granulated sugar
- 2½ teaspoons instant yeast
- ¾ teaspoon salt
- 2 large eggs, at room temperature
- ⅓ cup (75 milliliters) water
- ⅓ cup (5 tablespoons/ 71 grams) butter, softened
- 1 egg, beaten, for egg wash

FOR THE CHOCOLATE FILLING

- 1 cup (6 ounces/170 grams) chopped bittersweet chocolate
- 4 tablespoons (½ stick/ 57 grams) butter, softened
- ¼ cup (28 grams) confectioners' sugar, sifted
- ⅓ cup (37 grams) unsweetened cocoa powder

FOR THE SYRUP

- 3 tablespoons water
- 3 tablespoons sugar

1 **Make the dough:** In the bowl of a stand mixer fitted with the dough hook (or in a large bowl if kneading by hand), stir together the flour, sugar, yeast, and salt to combine.

2 In a small measuring cup, whisk together the eggs and water. With the stand mixer on medium speed, add this to the flour mixture and mix until the dough comes together and forms a ball, 3 to 4 minutes. (After a few minutes, if the dough looks a little dry, add a splash more water until the dough forms a mass and pulls away cleanly from the bowl.)

3 Turn down the mixer to low, add the butter a spoonful at a time and mix until incorporated. Mix on medium speed for roughly 8 minutes, until the dough is completely smooth. (If kneading by hand, knead the dough for 15 minutes.)

4 Lightly grease a bowl with oil and place the dough inside, cover with plastic wrap and a kitchen towel, and let proof at room temperature for roughly 2 hours. Transfer the bowl to the fridge and let the dough rest overnight.

5 **Make the chocolate filling:** Melt ½ cup (3 ounces/ 85 grams) of the chocolate and the butter together in a small saucepan over low heat (or in the microwave) until smooth. Stir in the confectioners' sugar and cocoa; the mixture should form a spreadable paste. Cover and store at room temperature overnight.

CONTINUES

BRAIDED CHOCOLATE BABKA

6   When you are ready to bake the babka, lightly grease a 9 x 5-inch loaf pan.

7   Turn the dough out onto a floured surface and roll it into a 10 x 14-inch rectangle, about ½ inch thick. Spread the chocolate mixture evenly over the dough, leaving a ½-inch border. (If the chocolate filling has firmed, you can soften it in the microwave for just a few seconds to make it spreadable.) Sprinkle the remaining ½ cup (3 ounces/85 grams) chopped chocolate evenly over the melted chocolate mixture.

8   Brush the exposed dough with the egg wash. Starting at one long side, roll the dough up into a long, tight cigar. Using a serrated knife, carefully cut the roll in half lengthwise and lay the halves next to each other, cut-sides up. Pinch the top ends of the strands together gently. Lift one side over the next, forming a twist and trying to keep the cut sides facing up. Don't worry if this step makes a mess, just do the best you can! Transfer the twist to the prepared loaf pan, still cut-side up.

9   Cover tightly with plastic wrap and let rise at room temperature for 60 to 90 minutes.

10   About 20 minutes before the end of the rising time, preheat the oven to 375°F (190°C).

11   Uncover the dough and bake for 30 to 35 minutes, until golden brown.

12   **Make the syrup:** In a small saucepan, combine the sugar and water and bring to a simmer, stirring until the sugar has dissolved. Remove from the heat and set aside to cool.

13   As soon as the babka is done baking and while it is still hot, brush it generously with all the syrup. Let it cool in the pan for 20 minutes, then transfer to a wire rack to cool completely before slicing and serving. Store leftovers in an airtight container at room temperature for up to 3 days.

CARAMEL PECAN MONKEY BREAD

This is a devilishly decadent sweet treat that is perfect to serve for brunch at home or on a day when everyone in the house needs a pick-me-up! It's designed to be pulled apart and shared, and I do have to remind myself of that when I'm eating it. Note that you'll make the dough and assemble the bread the day before, so the next morning it will be ready to bake and be eaten warm from the oven!

MAKES 1 BUNDT LOAF

FOR THE SWEET DOUGH

- 3¼ cups (461 grams) all-purpose flour
- 2 teaspoons instant yeast
- ¼ cup (57 grams) granulated sugar
- 2 teaspoons salt
- 1 cup (240 milliliters) whole milk
- ⅓ cup (71 milliliters) water
- 2 tablespoons (28 grams) butter, melted

FOR THE CARAMEL SAUCE

- ¾ cup (1½ sticks/ 170 grams) butter
- 1¾ cups (282 grams) dark brown sugar
- ½ cup (120 milliliters) heavy cream
- ⅛ teaspoon salt

FOR THE CINNAMON SUGAR

- ½ cup (115 grams) granulated sugar
- 2 tablespoons ground cinnamon

- 1 cup (142 grams) pecans, toasted and chopped

1 **Make the sweet dough:** In the bowl of a stand mixer fitted with the dough hook (or in a large bowl, if kneading by hand), stir together the flour, yeast, sugar, and salt.

2 In a large measuring cup, whisk together the milk, water, and melted butter. Add this to the flour mixture and mix on medium speed until the dough is shiny and smooth, 6 to 8 minutes. (If kneading by hand, knead for 15 minutes.) The dough should be soft and on the wet side.

3 Transfer the dough to an oiled bowl, turning it to coat lightly in the oil. Cover the bowl with plastic wrap and a kitchen towel and let the dough rise until doubled in size, 1½ to 2 hours.

4 **Make the caramel sauce:** In a small saucepan, combine the butter, brown sugar, cream, and salt and heat over medium-low heat until the sugar has dissolved and the butter has melted. Set aside to cool. (You can also do this step in a bowl in the microwave.)

5 Turn the dough out onto a floured surface and knead gently to deflate it. Using a bench scraper or knife, cut the dough into roughly equal-size pieces and roughly roll them into balls. (I don't give measurements here for the sizes of the pieces because there is no correct size—just make them roughly the same size.)

6 **Make the cinnamon sugar:** In a large bowl, mix together the sugar and cinnamon. Add the dough pieces and toss to coat each piece generously.

CONTINUES

CARAMEL PECAN MONKEY BREAD

7   **Assemble the monkey bread:** Drizzle one-third of the caramel sauce into a Bundt pan, then add one-third of the pecans. Layer half the dough pieces on top. Drizzle with half the remaining caramel sauce and sprinkle with half the remaining pecans. Top with the remaining dough pieces, followed by the remaining pecans and the last of the caramel sauce.

8   Cover the pan tightly with plastic wrap and refrigerate overnight where it will slowly proof.

9   The next morning, take the pan from the fridge and let the monkey bread rise at room temperature for 60 to 90 minutes until puffy and risen well.

10 About 20 minutes before the end of the rising time, preheat the oven to 350°F (180°C).

11 Uncover the pan and bake the monkey bread for 35 to 40 minutes, until the top is beautifully brown and the caramel begins to bubble around the edges. Let cool in the pan for 10 minutes (any longer and the bread will be too sticky and hard to remove!), then turn it out onto a platter or large plate and let cool slightly, about 10 minutes.

12 Enjoy straight away! Store leftovers in an airtight container at room temperature for up to 3 days.

MY GO-TO BRIOCHE LOAF

One of my most vivid memories from college was learning how to make brioche dough in pastry class. I had never made a dough with so much butter or so many eggs before. I distinctly remember baking it in the classic fluted tin and pushing in the little piece of dough on top, giving it that iconic brioche look. This recipe is a little more streamlined, but the soft, fluffy, and buttery results are exactly the same as I remember.

MAKES ONE LOAF

- 3 cups (426 grams) all-purpose flour
- ⅓ cup (71 grams) granulated sugar
- 1½ teaspoons instant yeast
- ½ teaspoon salt
- 1 cup (240 milliliters) whole milk, at room temperature
- 4 large egg yolks, at room temperature
- 2 tablespoons sour cream, at room temperature
- 6 tablespoons (¾ stick/ 85 grams) butter, softened
- 1 egg, beaten, for egg wash

1. In the bowl of a stand mixer fitted with the dough hook, stir together the flour, sugar, yeast, and salt.

2. In a measuring cup, whisk together the milk, egg yolks, and sour cream.

3. Add the wet ingredients to the dry ingredients and mix on medium-low speed for 3 to 4 minutes.

4. With the mixer running, gradually add the butter and mix for another 8 to 10 minutes. At this point, the dough should be smooth, elastic, and a little sticky.

5. Grease a large bowl with oil. Transfer the dough to the oiled bowl and turn the dough to coat with oil. Cover the bowl with plastic wrap and a kitchen towel and set aside in a warm place to proof for 1½ to 2 hours, until doubled in size.

6. Butter a 9 x 5-inch loaf pan.

7. After the dough has risen, turn it out onto a lightly floured surface and divide into 8 equal pieces. Working with one piece at a time, fold the edges toward the middle and then roll into a smooth ball. Place the balls of dough in the prepared loaf pan, seam-side down, in 4 rows of 2 balls each. Cover and let proof in a warm place until doubled in size, about 1 hour.

8. About 20 minutes before the dough is finished proofing, preheat the oven to 350°F (180°C).

9. Carefully brush the top of the loaf with the egg wash.

10. Bake for about 45 minutes, until the brioche is deep golden brown and sounds slightly hollow when tapped.

11. Let cool completely before slicing and serving. Store leftovers in an airtight container at room temperature for up to 3 days. This loaf also freezes beautifully for up to 4 weeks.

STICKY MAPLE WALNUT MORNING BUNS

This sticky maple and walnut swirl Danish is why they invented brunching! Its flaky, buttery dough is crammed with toasted nuts and cinnamon sugar that caramelizes at the bottom to create a beautiful caramel on the outside. Good luck stopping at just one serving because I couldn't! Just be sure to plan ahead: you'll need to refrigerate the dough overnight before assembling and baking the buns the next day.

MAKES 12 BUNS

1 recipe Danish Dough (page 228)
Melted butter, for greasing

FOR THE FILLING

½ cup (1 stick/115 grams) butter
1 cup (170 grams) dark brown sugar
¼ cup (71 grams) pure maple syrup
1 tablespoon ground cinnamon
½ teaspoon salt
1½ cups (213 grams) walnuts, toasted and finely chopped

FOR THE TOPPING

¼ cup (71 grams) pure maple syrup
4 tablespoons (½ stick/ 57 grams) butter

1 Make the Danish dough and refrigerate for at least 12 hours or up to overnight. When the dough is chilled, generously butter a 12-cup muffin tin.

2 **Make the filling:** In a small saucepan, stir together the butter, brown sugar, maple syrup, cinnamon, and salt and heat over medium heat, stirring, until the butter and sugar have melted. Remove from the heat, stir in the walnuts, and let cool.

3 On a floured surface, roll out the Danish dough to a 12 x 18-inch rectangle. Spread the cooled filling evenly over the dough, leaving a 1-inch border all the way around.

4 Starting with a long edge, gently roll up the dough into a log. With a serrated knife, slice the roll into 2-inch-thick pieces and place one slice in each well of the prepared muffin tin, cut-side up. Cover the dough and let rise in a warm place until puffed up, 45 to 60 minutes.

5 About 20 minutes before the end of the rising time, preheat the oven to 375°F (190°C).

6 Carefully uncover the dough.

7 Bake for 20 to 25 minutes, until golden brown.

8 **Make the topping:** Warm the maple syrup and butter together until the butter melts. Set aside.

9 Remove the buns from the oven, leaving them in the pan. Generously brush the warm buns with the maple-butter topping. Let cool for 15 minutes, then carefully remove the buns from the pan and place them on a baking sheet to cool completely. Spoon any walnut filling that has spilled out back into the buns.

10 These buns are best eaten the day they are made, but you can store any leftovers in an airtight container at room temperature overnight. Rewarm in the microwave, or in a 300°F (150°C) oven for 10 minutes.

ALMOND TWIST

One of the hardest parts of making a book is choosing which recipes to include, but this almond breakfast ring was too good to leave out. My first attempt didn't look the part, but it tasted *amazing*! If you know me, you know I can't leave well enough alone, so I went back to the drawing board and tested this recipe again and again until I got the look to match that fantastic taste. (And filled it with as many almonds as I could!) More testing means more tasting, which you'll never find me complaining about.

MAKES 12 SERVINGS

1 recipe Danish Dough (page 228)

1½ cups (170 grams) almond flour

¾ cup (170 grams) granulated sugar

6 tablespoons (¾ stick/ 85 grams) butter, softened

3 tablespoons all-purpose flour

1 large egg

1½ teaspoons pure almond extract

1 egg, beaten, for egg wash

¼ cup (35 grams) sliced almonds

1 Make the Danish dough and refrigerate for at least 12 hours or up to overnight. When the dough is chilled, butter a 10-inch springform pan.

2 In a medium bowl, combine the almond flour, sugar, butter, all-purpose flour, egg, and almond extract and stir until well mixed. Place in the fridge to firm up for a minimum of 1 hour.

3 On a floured surface, roll out the Danish dough into a 12 x 18-inch rectangle. Spread the almond filling evenly over the dough.

4 Starting with one long edge, roll up the dough into a log.

5 Using a serrated knife, carefully cut the roll in half lengthwise and lay both halves next to each other, cut sides up. Pinch the top ends gently together. Lift one side over the next, forming a twist, keeping the cut sides facing up. Press both ends of the twist together to form a ring. Place in your prepared springform pan. (See Braided Chocolate Babka photos on page 168 for reference.)

6 Cover well with plastic wrap and a kitchen towel and proof at room temperature for 1 hour, or until well risen.

7 About 20 minutes before the end of the rising time, preheat the oven to 375°F (190°C).

8 Carefully brush the ring all over with the egg wash and then sprinkle with the sliced almonds.

9 Bake for 35 to 45 minutes, until the pastry is a deep golden brown. Let cool slightly in the pan before serving.

10 Slice and serve while still warm. Store leftovers, covered, at room temperature for up to 2 days.

BUTTERY FRUIT DANISHES

Pour yourself a dark cup of coffee, grab the paper, and enjoy a plate of these addicting fruit Danishes! These breakfast treats are incredibly flaky, with the perfect combination of fruit and jam. Prepare them the night before so you can have a little slice of heaven any day of the week.

MAKES 12 DANISHES

- 1 recipe Danish Dough (page 228)
- 2 medium Granny Smith apples (about 400 grams), peeled, cored, and finely chopped
- 1 cup (142 grams) raisins
- 2 tablespoons dark brown sugar
- 1 teaspoon ground cinnamon
- 1 cup (282 grams) apricot jam
- 1 egg, beaten, for egg wash

FOR THE VANILLA GLAZE

- 1/3 cup (37 grams) confectioners' sugar
- 1 teaspoon whole milk
- 1 teaspoon pure vanilla extract

1 Make the Danish dough and refrigerate for at least 12 hours or up to overnight. When the dough is chilled, line two baking sheets with parchment paper.

2 In a medium bowl, combine the apples, raisins, brown sugar, and cinnamon.

3 On a floured surface, roll out the dough to a 12 x 18-inch rectangle. Spread the apricot jam over the dough, leaving a 1-inch border, then sprinkle the apple mixture evenly over the jam.

4 Starting with one long side, roll up the dough into a log.

5 Slice the log of dough into twelve 1-inch-thick rounds and place them cut-side up on the prepared baking sheets. Cover loosely with plastic wrap and a kitchen towel and let rise until doubled, 30 to 60 minutes.

6 About 20 minutes before the end of the rising time, preheat the oven to 375°F (190°C). Carefully brush the pastry with the egg wash.

7 Bake for 15 to 20 minutes, until golden brown, then let cool completely on the baking sheet.

8 **Make the glaze:** In a small bowl, whisk together the confectioners' sugar, milk, and vanilla until smooth.

9 Drizzle the glaze over the cooled pastries and let set before serving. These are best enjoyed the day they are baked, but leftovers can be stored in an airtight container for up to 1 day.

HAWAIIAN SWEET ROLLS

You can never have too many bread doughs in your repertoire. This dough is as light as air and *so* versatile! Use these slightly sweet rolls for a sandwich or even French toast—whatever you choose, not a roll will go to waste.

MAKES 20 ROLLS

5½ cups (781 grams) all-purpose flour

1 tablespoon instant yeast

2½ teaspoons salt

1 large egg, at room temperature

1 cup (240 milliliters) pineapple juice

½ cup (120 milliliters) whole milk

6 tablespoons (¾ stick/ 85 grams) butter, softened

⅓ cup (115 grams) honey

2 teaspoons white vinegar

2 teaspoons pure vanilla extract

1 Butter a 9 x 13-inch baking dish. Set aside.

2 In the bowl of a stand mixer fitted with the dough hook (or in a large bowl, if kneading by hand), combine the flour, yeast, and salt.

3 In a medium bowl, whisk together the egg, pineapple juice, milk, butter, honey, vinegar, and vanilla.

4 Add the wet ingredients to the dry ingredients and mix on low speed until the flour is moistened, then increase the speed to medium-low and knead for 8 to 10 minutes, until the dough is smooth and elastic but still slightly sticky. (If kneading by hand, knead for 12 to 15 minutes.)

5 Transfer the dough to a large, greased bowl, cover with plastic wrap and a kitchen towel, and let rise in a warm place until doubled in size, about 1½ hours.

6 Turn the dough out onto a floured surface and press down gently to deflate it, then divide the dough into 20 equal-size portions. Roll each portion of dough into a smooth ball and arrange the balls in the prepared baking dish in 5 rows of 4 balls each. Cover again and let rise until doubled again, 1 to 1½ hours.

7 About 20 minutes before the end of the rising time, preheat the oven to 375°F (190°C).

8 Uncover the rolls. Bake for 25 to 30 minutes, until deep golden brown.

9 Let cool for 20 minutes before serving. Store leftovers in an airtight container at room temperature for up to 3 days. They also freeze well for up to 4 weeks.

Chapter 7

SHORT *and* SWEET ANY DAY

Here is a collection of recipes that give you maximum results for minimum work! They are perfect for any day of the week when you need something fast, homemade, and delicious. These short-and-sweet desserts also don't require a lot of ingredients.

With the help of the Master Recipes, you can get a little creative with one crisp topping or pie crust and make a multitude of desserts! So get ready to have some fun putting your own twist on these classics.

FRUIT FOOLS

Sometimes the simplest recipes have the most satisfying results—and it doesn't get much simpler than some whipped cream and some of your favorite fruit!

STRAWBERRY and MINT FOOL

MAKES 6 SERVINGS

- 2 cups (10 ounces/ 282 grams) fresh strawberries, halved
- 1 tablespoon water
- 1 tablespoon granulated sugar
- 1 tablespoon chopped fresh mint
- 1 recipe Whipped Cream (page 232)

1 In a small microwave-safe bowl, mix together the strawberries, water, sugar, and mint.

2 Heat in the microwave just to warm through and soften the berries, getting them to release their juices, about 60 seconds. (Alternatively, combine the ingredients in a small saucepan and warm on the stovetop.) Set aside to cool.

3 When ready to serve, mix the macerated fruit with the whipped cream and divide among six individual serving dishes. Enjoy immediately!

BLUEBERRY and ROSE WATER FOOL

MAKES 6 SERVINGS

- 2 cups (10 ounces/ 282 grams) fresh blueberries
- 1 tablespoon water
- 1 tablespoon granulated sugar
- 1 teaspoon rose water
- 1 recipe Whipped Cream (page 232)

1 In a small microwave-safe bowl, mix together the blueberries, water, sugar, and rose water.

2 Heat in the microwave just to warm through and soften the berries, getting them to release their juices, about 60 seconds. (Alternatively, combine the ingredients in a small saucepan and warm on the stovetop.) Set aside to cool.

3 When ready to serve, mix the macerated fruit with the whipped cream and divide among six individual serving dishes. Enjoy immediately!

Strawberry and
Mint Fool

Blueberry and
Rose Water Fool

Boozy
Raspberry Fool

Summer
Blackberry Fool

BOOZY RASPBERRY FOOL

2 cups (10 ounces/
   282 grams) fresh
   raspberries

1 tablespoon water

1 tablespoon
   granulated sugar

1 tablespoon Chambord

1 recipe Whipped Cream
   (page 232)

1 In a small microwave-safe bowl, mix together the raspberries, water, sugar, and Chambord.

2 Heat in the microwave just to warm through and soften the berries, getting them to release their juices, about 60 seconds. (Alternatively, combine the ingredients in a small saucepan and warm on the stovetop.) Set aside to cool.

3 When ready to serve, mix the macerated fruit with the whipped cream and divide among six individual serving dishes. Enjoy immediately!

SUMMER BLACKBERRY FOOL

2 cups (10 ounces/
   282 grams) fresh
   blackberries

1 tablespoon water

1 tablespoon
   granulated sugar

1 tablespoon crème
   de cassis

1 recipe Whipped Cream
   (page 232)

1 In a small microwave-safe bowl, mix together the blackberries, water, sugar, and crème de cassis.

2 Heat in the microwave just to warm through and soften the berries, getting them to release their juices, about 60 seconds. (Alternatively, combine the ingredients in a small saucepan and warm on the stovetop.) Set aside to cool.

3 When ready to serve, mix the macerated fruit with the whipped cream and divide among six individual serving dishes. Enjoy immediately!

FRUIT CRISPS

A fruit crisp is and always will be one of my all-time favorite desserts! In fact, I made sure to test *multiple* crisp recipes to get the exact results that I wanted. These are the things that keep me up at night! But I'm happy to say that this fruit crisp has the perfect flavor, texture, and ratio of fruit to topping. Use fresh or frozen fruit for these recipes.

TRIPLE BERRY CRISP

MAKES 6 SERVINGS

- 1 recipe Crisp Topping (page 226)
- 6 cups (30 ounces/ 852 grams) mixed summer berries
- 2½ tablespoons light brown sugar
- 1 tablespoon cornstarch
- 1 tablespoon fresh lemon juice

  Vanilla ice cream, for serving

1 Preheat the oven to 375°F (190°C).

2 Make the crisp topping and refrigerate until needed.

3 In a large bowl, combine the berries, brown sugar, cornstarch, and lemon juice.

4 Transfer the berry mixture to a round 10-inch baking dish and sprinkle the crisp topping evenly over the berries.

5 Bake for 50 to 60 minutes, until the topping is golden brown and the fruit is bubbling.

6 Serve warm, with vanilla ice cream. Store leftovers, covered, at room temperature for up to 2 days.

TRADITIONAL APPLE CRISP

MAKES 6 SERVINGS

- 1 recipe Crisp Topping (page 226)
- 6 medium Granny Smith apples (about 30 ounces/852 grams), peeled and chopped
- 2½ tablespoons dark brown sugar
- 1 tablespoon cornstarch
- 1 tablespoon fresh lemon juice

  Vanilla ice cream, for serving

1 Preheat the oven to 375°F (190°C).

2 Make the crisp topping and refrigerate until needed.

3 In a large bowl, combine the apples, brown sugar, cornstarch, and lemon juice.

4 Transfer the apple mixture to a round 10-inch baking dish and sprinkle the crisp topping evenly over the apples.

5 Bake for 50 to 60 minutes, until the topping is golden brown and the fruit is bubbling.

6 Serve warm, with vanilla ice cream. Store leftovers, covered, at room temperature for up to 2 days.

Triple Berry Crisp

Traditional Apple Crisp

Blueberry Peach Crisp

Rhubarb and Strawberry Crisp

BLUEBERRY PEACH CRISP

1   recipe Crisp Topping (page 226)

4   cups (20 ounces/ 568 grams) pitted and chopped peaches (roughly 1-inch pieces)

2   cups (10 ounces/ 282 grams) blueberries

2½ tablespoons light brown sugar

1   tablespoon cornstarch

1   tablespoon fresh lemon juice

   Vanilla ice cream, for serving

1 Preheat the oven to 375°F (190°C).

2 Make the crisp topping and refrigerate until needed.

3 In a large bowl, combine the peaches, blueberries, brown sugar, cornstarch, and lemon juice.

4 Transfer the fruit mixture to a round 10-inch baking dish and sprinkle the crisp topping evenly over the fruit.

5 Bake for 50 to 60 minutes, until the topping is golden brown and the fruit is bubbling.

6 Serve warm, with vanilla ice cream. Store leftovers, covered, at room temperature for up to 2 days.

RHUBARB *and* STRAWBERRY CRISP

1   recipe Crisp Topping (page 226)

3   cups (15 ounces/ 426 grams) sliced rhubarb (roughly 1-inch pieces)

3   cups (15 ounces/ 426 grams) hulled strawberries, halved

2½ tablespoons light brown sugar

1   tablespoon cornstarch

1   tablespoon lemon juice

   Vanilla ice cream, for serving

1 Preheat the oven to 375°F (190°C).

2 Make the crisp topping and refrigerate until needed.

3 In a large bowl, combine the rhubarb, strawberries, brown sugar, cornstarch, and lemon juice.

4 Transfer the fruit mixture to a round 10-inch baking dish and sprinkle the crisp topping evenly over the fruit.

5 Bake for 50 to 60 minutes, until the topping is golden brown and the fruit is bubbling.

6 Serve warm, with vanilla ice cream. Store leftovers, covered, at room temperature for up to 2 days.

CLAFOUTIS

A baked flan for all the world (as we would say in Ireland!), this simple but sophisticated dessert lends itself beautifully to being baked with a variety of seasonal fruits. As soon as summer rolls around and cherries hit the shelves, load up to make this quick dessert.

CLAFOUTI BATTER

This batter is a perfect blank canvas—pour it over any in-season fruit, bake, and you'll have a beautiful, traditional French custard. Get creative and try one of my variety of clafoutis!

⅔ cup (94 grams) all-purpose flour

½ cup (115 grams) granulated sugar

¼ teaspoon salt

¾ cup (180 milliliters) whole milk

3 large eggs

2 tablespoons (28 grams) butter, melted

1 teaspoon pure vanilla extract

1 In a large bowl, whisk together the flour, sugar, and salt.

2 In a medium bowl, stir together the milk, eggs, butter, and vanilla.

3 Add the wet ingredients to the dry ingredients and whisk until you have a smooth batter. This batter is now ready to use in any of my clafouti recipes (see pages 199 to 201) or in your own version. Go to those recipes for baking times and temperatures.

CHERRY AMARETTO CLAFOUTIS

MAKES 5 SERVINGS

1 recipe Clafouti Batter (page 198)

4 cups (20 ounces/ 568 grams) fresh cherries, pitted

2 tablespoons amaretto liqueur

Whipped Cream (page 232), for serving

1 Make the clafouti batter and refrigerate until needed.

2 In a medium bowl, combine the cherries and amaretto and allow to macerate for at least 30 minutes.

3 Preheat the oven to 400°F (200°C). Butter five 4-ounce baking dishes and place them on a baking sheet.

4 Divide the fruit and accumulated juices among the prepared baking dishes.

5 Pour the clafouti batter on top of the cherries, filling the dishes almost to the top.

6 Bake for 15 minutes, then reduce the oven temperature to 350°F (180°C) and bake for another 20 to 25 minutes, until the clafoutis are firm and baked through.

7 Serve warm, with whipped cream. Store leftovers, covered, in the fridge for up to 2 days.

PEACH and RASPBERRY CLAFOUTIS

MAKES 6 SERVINGS

1 recipe Clafouti Batter (page 198)

2 cups (10 ounces/ 284 grams) pitted and sliced peaches

1 cup (5 ounces/142 grams) raspberries

Whipped Cream (page 232), for serving

1 Make the clafouti batter and refrigerate until needed.

2 Preheat the oven to 400°F (200°C). Butter a 9-inch baking dish and place on a baking sheet.

3 In a medium bowl, combine the peaches and raspberries, then pour the fruit into the prepared baking dish.

4 Pour the clafouti batter on top of the fruit, filling the dish almost to the top.

5 Bake for 15 minutes, then reduce the oven temperature to 350°F (180°C) and bake for another 25 to 35 minutes, until the clafouti is firm and baked through.

6 Serve warm, with whipped cream. Store leftovers, covered, in the fridge for up to 2 days.

Cherry Amaretto
Clafoutis

Peach and Raspberry Clafoutis

Blackberry Clafoutis

Fresh Fig and Cardamom Clafoutis

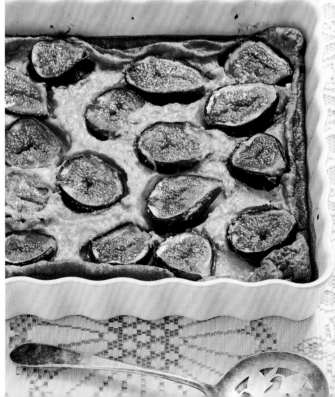

BLACKBERRY CLAFOUTIS

1 recipe Clafouti Batter (page 198)

2½ cups (12 ounces/ 340 grams) blackberries

Whipped Cream (page 232), for serving

1 Make the clafouti batter and refrigerate until needed.

2 Preheat the oven to 400°F (200°C). Butter a 9-inch baking dish and place it on a baking sheet.

3 Arrange the blackberries in the baking dish.

4 Pour the clafouti batter on top of the blackberries, filling the dish almost to the top.

5 Bake for 15 minutes, then reduce the oven temperature to 350°F (180°C) and bake for another 25 to 35 minutes, until the clafouti is firm and baked through.

6 Serve warm, with whipped cream. Store leftovers, covered, in the fridge for up to 2 days.

FRESH FIG *and* CARDAMOM CLAFOUTIS

1 recipe Clafouti Batter (page 198)

3¼ cups (1 pound/450 grams) fresh figs, halved

¼ teaspoon ground cardamom

Whipped Cream (page 232), for serving

1 Make the clafouti batter and refrigerate until needed.

2 Preheat the oven to 400°F (200°C). Butter a 9-inch baking dish and place it on a baking sheet.

3 Arrange the figs in the baking dish.

4 Whisk the ground cardamom into the clafouti batter, then pour the batter over the figs, filling the dish almost to the top.

5 Bake for 15 minutes, then reduce the oven temperature to 350°F (180°C) and bake for another 25 to 35 minutes, until the clafouti is firm and baked through.

6 Serve warm, with whipped cream. Store leftovers, covered, in the fridge for up to 2 days.

PAN-ROASTED FRUIT

Don't overthink it! Grab a pan and some fresh fruit, and create these easy but impressive desserts. The flavors just sing together in the pan.

ROASTED PINEAPPLE *with* RUM COCONUT SAUCE

MAKES 4 SERVINGS

- 1 large pineapple, peeled, cored, and sliced ½ inch thick
- ½ cup (85 grams) dark brown sugar
- ¼ cup (57 milliliters) spiced rum or dark rum
- 2 tablespoons (28 grams) butter
- ½ cup (120 milliliters) full-fat coconut milk
- Vanilla ice cream, for serving

1 In a large bowl, toss the pineapple pieces with the brown sugar and rum.

2 Heat a large skillet over medium heat and when hot, add the butter to the pan and lay down as many pineapple pieces as you can fit in a single layer. Cook the pieces until deeply caramelized, 3 to 4 minutes, then flip them and cook the other side for another 2 minutes.

3 As the pineapple slices are done, divide them among six individual serving dishes and repeat with the remaining pineapple.

4 Pour any remaining rum and sugar left in the bowl into the pan, add the coconut milk, and simmer for 2 to 3 minutes to reduce and thicken your sauce.

5 Pour the coconut sauce over the pineapple pieces, top each dish with a scoop of vanilla ice cream, and serve immediately.

Roasted Pineapple with
Rum Coconut Sauce

Roasted
Strawberries
with
Rose and
Cardamom
Yogurt

Honey Glazed Figs with
Whipped Mascarpone

Caramelized
Bananas with
Rosemary

ROASTED STRAWBERRIES *with* ROSE *and* CARDAMOM YOGURT

MAKES 4 SERVINGS

- 1¼ cups (282 grams) plain Greek yogurt
- 4 tablespoons (57 grams) granulated sugar
- ½ teaspoon ground cardamom
- 4 cups (20 ounces/ 568 grams) fresh strawberries, hulled and halved
- 1 tablespoon rose water

1 In a small bowl, stir together the yogurt, 2 tablespoons of the sugar, and the cardamom. Set aside for 15 minutes to allow the sugar to dissolve.

2 In a large skillet, combine the strawberries and remaining 2 tablespoons sugar and cook over medium heat for 3 to 4 minutes, until the strawberries have softened slightly and released their juices and the sugar has dissolved. Remove from the heat and stir in the rose water. Divide the strawberries among four individual serving bowls.

3 Give the yogurt another quick stir, then dollop it on top of the strawberries and serve immediately.

HONEY GLAZED FIGS *with* WHIPPED MASCARPONE

MAKES 4 SERVINGS

1 cup (225 grams) mascarpone cheese, at room temperature

2 tablespoons plus ⅓ cup (115 grams) honey

Grated zest of 1 lemon

½ cup (1 stick/115 grams) butter

18 figs, halved lengthwise

1 Whisk the mascarpone with the 2 tablespoons of honey and the lemon zest until combined. Place in the fridge until needed.

2 In a large skillet, combine the butter and the remaining ⅓ cup (115 grams) honey and heat over medium heat until the mixture bubbles. Add the figs, cut-side down, and cook for about 2 minutes, until caramelized. Flip and cook on the other side for 2 minutes, or until softened.

3 Spread the mascarpone on your serving platter and arrange the figs on top. Drizzle with any sauce left in the pan and serve immediately.

CARAMELIZED BANANAS *with* ROSEMARY

MAKES 4 SERVINGS

1 cup (170 grams) dark brown sugar

½ cup (1 stick/115 grams) butter

¼ cup (60 milliliters) water

1 (8-inch) sprig rosemary

5 medium bananas (about 12½ ounces/354 grams), peeled and sliced into 1-inch-thick rounds

Vanilla ice cream, for serving

1 In a large skillet, combine the brown sugar, butter, and water and heat over medium heat, stirring, until the sugar has dissolved. Stir in the rosemary and continue to cook until the mixture begins to caramelize.

2 Add the banana slices and cook for 2 to 3 minutes, until browned, then flip and cook on the other side for 2 minutes. Remove from the heat.

3 Discard the sprig of rosemary and divide the bananas and caramel sauce among six serving bowls. Top with a scoop of vanilla ice cream and serve immediately.

CROSTATAS

In my opinion, crostata is one of the most beautiful desserts that takes very little effort. It's the perfect way to take advantage of any fruit in season—or you can never go wrong with chocolate and nuts!

CHOCOLATE, HAZELNUT, *and* MASCARPONE CROSTATA

MAKES 6 SERVINGS

- 1 cup (141 grams) hazelnuts, toasted
- ⅔ cup (4 ounces/115 grams) finely chopped bittersweet chocolate
- 1 cup (225 grams) mascarpone cheese, at room temperature
- ⅓ cup (71 grams) granulated sugar
- 1 large egg, at room temperature
- 1 large egg white, at room temperature
- 1 tablespoon hazelnut liqueur (optional)
- 1 recipe Pie Crust (page 222)
- 1 egg, beaten, for egg wash
  Vanilla ice cream, for serving

1 Preheat the oven to 375°F (190°C). Line a baking sheet with parchment paper.

2 In a food processor, finely grind ¾ cup (106 grams) of the hazelnuts.

3 Melt the chocolate in a double boiler or in the microwave. Add the melted chocolate, mascarpone, sugar, egg, egg white, and hazelnut liqueur (if using) to the food processor and process to combine. Set aside.

4 On a floured surface, roll out your pie dough into a 12-inch round, about ⅛ inch thick, and transfer it to the prepared baking sheet.

5 Pour the chocolate mixture into the center of the dough, leaving a 2-inch border all around. Fold the pie crust border over the chocolate mixture, allowing the edges to randomly pleat and leaving the filling exposed in the center. Very gently press down on the pleats to seal, then brush the dough with the egg wash.

6 Coarsely chop the remaining ¼ cup (35 grams) hazelnuts and sprinkle them over the exposed chocolate filling.

7 To prevent spreading while baking, you can remove the bottom from a 9-inch springform pan and place just the ring on the baking sheet around the crostata to help the pastry hold its shape while baking. This step is optional.

8 Bake for 45 to 50 minutes, until the crust is golden brown and the filling is set.

9 Let cool before serving with vanilla ice cream. Store leftovers, loosely covered, at room temperature for up to 2 days.

CARAMEL APPLE CROSTATA

MAKES 6 SERVINGS

3 medium Granny Smith apples (about 15 ounces/ 426 grams), peeled, cored, and thinly sliced

1 tablespoon fresh lemon juice

1 teaspoon ground cinnamon

¼ cup (57 grams) Honey Toffee Sauce (page 230)

1 recipe Pie Crust (page 222)

1 egg, beaten, for egg wash

Granulated sugar, for sprinkling

2 tablespoons (28 grams) butter, diced

Vanilla ice cream, for serving

1 Preheat the oven to 375°F (190°C). Line a baking sheet with parchment paper.

2 In a large bowl, toss the apples with the lemon juice, cinnamon, and honey toffee sauce.

3 On a floured surface, roll out the pie dough into a 12-inch round, about ⅛ inch thick, and transfer it to the prepared baking sheet.

4 Arrange the apples over the dough round, leaving a 2-inch border. Fold the exposed dough up over the apples, allowing the pastry to randomly pleat and leaving the apples exposed in the center. Gently press down on the pleats to seal, then brush the dough with the egg wash and sprinkle with sugar. Finally, dot the exposed apples with the butter.

5 To prevent spreading while baking, you can remove the bottom from a 9-inch springform pan and place just the ring on the baking sheet around the crostata to help the pastry hold its shape while baking. This step is optional.

6 Bake for 45 to 50 minutes, until the apples are tender and the crust is golden brown.

7 Let cool before serving with vanilla ice cream. Store leftovers, loosely covered, at room temperature for up to 2 days.

Chocolate, Hazelnut, and Mascarpone Crostata

Caramel Apple Crostata

Berry and Cream Cheese Crostata

Plum, Crème Fraîche, and Vanilla Bean Crostata

BERRY and CREAM CHEESE CROSTATA

MAKES 6 SERVINGS

1¼ cups (6¼ ounces/ 177 grams) mixed berries, sliced if large

2 teaspoons fresh lemon juice

1 teaspoon cornstarch

½ cup (115 grams) plus 2 teaspoons granulated sugar, plus more for sprinkling

1 cup (8 ounces/225 grams) cream cheese, at room temperature

2 teaspoons grated lemon zest

1 large egg, at room temperature

½ teaspoon pure vanilla extract

1 recipe Pie Crust (page 222)

1 egg, beaten, for egg wash

Vanilla ice cream, for serving

1 Preheat the oven to 375°F (190°C). Line a baking sheet with parchment paper.

2 In a small bowl, combine the berries, lemon juice, cornstarch, and 2 teaspoons of the sugar. Set aside.

3 In a medium bowl, whisk together the cream cheese, remaining ½ cup (115 grams) sugar, and the lemon zest until fully combined. Whisk in the egg and vanilla and set aside.

4 On a floured surface, roll out the pie dough into a 12-inch round, about ⅛ inch thick, and transfer it to the prepared baking sheet.

5 Pour the cream cheese filling into the center of the dough round and spread it evenly, leaving a 2-inch border. Scatter the berry mixture over the cream cheese. Fold the exposed dough up over the filling, allowing the pastry to randomly pleat and leaving the filling exposed in the center. Very gently press down on the pleats to seal, then brush the dough with the egg wash and sprinkle with some sugar.

6 To prevent spreading while baking, you can remove the bottom from a 9-inch springform pan and place just the ring on the baking sheet around the crostata to help the pastry hold its shape while baking. This step is optional.

7 Bake for 45 to 50 minutes, until the filling is set and the crust is golden brown.

8 Let cool before serving with vanilla ice cream. Store leftovers, loosely covered, at room temperature for up to 2 days.

PLUM, CRÈME FRAÎCHE, *and* VANILLA BEAN CROSTATA

MAKES 6 SERVINGS

3 cups (15 ounces/ 426 grams) pitted and sliced plums

2 tablespoons dark brown sugar

1 vanilla bean, split lengthwise and seeds scraped out

⅛ teaspoon salt

½ cup (115 grams) crème fraîche

2 tablespoons granulated sugar, plus more for sprinkling

1 recipe Pie Crust (page 222)

1 egg, beaten, for egg wash

2 tablespoons (28 grams) butter, diced

Vanilla ice cream, for serving

1 Preheat the oven to 375°F (190°C). Line a baking sheet with parchment paper.

2 In a medium bowl, toss together the plum slices, brown sugar, vanilla seeds, and salt. Set aside.

3 In a small bowl, combine the crème fraîche and granulated sugar.

4 On a floured surface, roll out the pie dough into a 12-inch round, about ⅛ inch thick, and transfer it to the prepared baking sheet.

5 Pour the crème fraîche mixture into the center of the dough round and spread it evenly, leaving a 2-inch border. Arrange the plums on top of the crème fraîche and drizzle with the juices accumulated at the bottom of the bowl. Fold the exposed dough up over the filling, allowing the pastry to randomly pleat and leaving the filling exposed in the center. Very gently press down on the pleats to seal, then brush with the egg wash and sprinkle with some sugar. Finally, dot the butter on top of the plums.

6 To prevent spreading while baking, you can remove the bottom from a 9-inch springform pan and place just the ring on the baking sheet around the crostata to help the pastry hold its shape while baking. This step is optional.

7 Bake for 45 to 50 minutes, until the filling is set and the crust is golden brown.

8 Let cool before serving with vanilla ice cream. Store leftovers, loosely covered, at room temperature for up to 2 days.

UPSIDE-DOWN CAKES

This is my type of comfort food! What could be better than sweet, juicy fruit and a pillowy soft cake? I request that you always serve it warm, and a healthy serving of whipped cream for topping is nonnegotiable!

I wanted to show how one batter can create an array of desserts, all unique in flavor, just by adding your favorite fruit and a few other ingredients. Find a variety of upside-down cakes on pages 212 to 216.

UPSIDE-DOWN CAKE BATTER

MAKES 8 SERVINGS

- ½ cup (1 stick/115 grams) butter, softened
- 1½ cups (340 grams) granulated sugar
- 3 large eggs, at room temperature
- 1 cup (240 milliliters) whole milk
- 1 teaspoon pure vanilla extract
- 1¼ cups (177 grams) all-purpose flour
- ½ cup (57 grams) almond flour
- 2 teaspoons baking powder
- ½ teaspoon salt

1 In the bowl of a stand mixer fitted with the paddle attachment or in a large bowl using a handheld mixer, beat the butter and sugar on high speed until light and fluffy, 3 to 4 minutes.

2 Add the eggs one at a time, followed by the milk and vanilla.

3 In a small bowl, whisk together the all-purpose flour, almond flour, baking powder, and salt, then fold the dry ingredients into the wet ingredients until combined.

4 Use the batter as directed in the individual recipes, following the indicated baking times and temperatures.

MAPLE and PEAR UPSIDE-DOWN CAKE

MAKES 8 SERVINGS

1   recipe Upside-Down
    Cake Batter (page 211)

¾   cup (213 milliliters) pure
    maple syrup

¼   cup (43 grams) dark
    brown sugar

4   tablespoons (½ stick/
    57 grams) butter

3   ripe but firm large pears
    (15 ounces/426 grams),
    peeled, cored, and sliced
    ¼ inch thick

1   Make the cake batter and refrigerate until
    ready to use.

2   Preheat the oven to 350°F (180°C). Butter a 9-inch
    round cake pan.

3   In a small saucepan, combine the maple syrup, brown
    sugar, and butter and heat over medium heat for 1 to
    2 minutes, until you have a golden brown sauce. Pour
    into the prepared baking pan.

4   Arrange the pear slices on top of the maple sauce so
    that the sauce is completely covered by the fruit. Pour
    the cake batter on top.

5   Bake for 50 to 60 minutes, until a toothpick inserted
    into the center comes out clean. While the cake
    is still hot, run a thin knife around the edge of the
    cake to loosen it from the pan, then invert it onto a
    serving plate.

6   Let cool slightly before serving. Store leftovers in
    an airtight container at room temperature for up
    to 2 days.

Maple and Pear Upside-Down Cake

Orange and Honey Toffee
Upside-Down Cake

Banana and Rum Upside-Down Cake

Fall Apple
Cinnamon
Upside-Down
Cake

ORANGE *and* HONEY TOFFEE UPSIDE-DOWN CAKE

MAKES 8 SERVINGS

1   recipe Upside-Down Cake Batter (page 211)

½   cup (142 grams) Honey Toffee Sauce (page 230)

4 or 5   medium navel oranges (15 ounces/426 grams)

1   Make the cake batter and refrigerate until ready to use.

2   Preheat the oven to 350°F (180°C). Butter a 9-inch round cake pan.

3   With a sharp knife, cut the peel off the navel oranges and then slice them into ½-inch-thick rounds.

4   Spread the honey toffee sauce in the prepared pan. Arrange the orange slices on top of the toffee sauce so the sauce is completely covered by the oranges. Pour the cake batter on top.

5   Bake for 50 to 60 minutes, until a wooden skewer inserted into the center comes out clean. While the cake is still hot, run a thin knife around the edge of the cake to loosen it from the pan, then invert it onto a serving plate.

6   Let cool slightly before serving. Store leftovers in an airtight container at room temperature for up to 2 days.

BANANA *and* RUM UPSIDE-DOWN CAKE

MAKES 8 SERVINGS

1 recipe Upside-Down Cake Batter (page 211)

⅓ cup (75 milliliters) spiced rum or dark rum

½ cup (85 grams) dark brown sugar

4 tablespoons (½ stick/ 57 grams) butter

4 medium bananas (roughly 15 ounces/426 grams)

1 Make the cake batter and refrigerate until ready to use.

2 Preheat the oven to 350°F (180°C). Butter a 9-inch cake pan.

3 In a small saucepan, combine the rum, brown sugar, and butter and bring to a simmer over medium-low heat. Simmer for 4 to 5 minutes, until you have a golden brown sauce. Pour the sauce into the prepared pan.

4 Peel the bananas and carefully slice them lengthwise. Arrange the bananas on top of the rum sauce so the sauce is completely covered by the bananas. Pour the cake batter on top.

5 Bake for 50 to 60 minutes, until a toothpick inserted into the center comes out clean. While the cake is still hot, run a thin knife around the edge of the cake to loosen it from the pan, then invert it onto a serving plate.

6 Let cool slightly before serving. Store leftovers in an airtight container at room temperature for up to 2 days.

FALL APPLE CINNAMON UPSIDE-DOWN CAKE

MAKES 8 SERVINGS

1   recipe Upside-Down Cake Batter (page 211)

½   cup (85 grams) dark brown sugar

4   tablespoons (½ stick/ 57 grams) butter

2   tablespoons heavy cream

1   teaspoon ground cinnamon

3   medium Granny Smith apples (roughly 15 ounces/426 grams), peeled, cored, and sliced ¼ inch thick

1   Make the cake batter and refrigerate until ready to use.

2   Preheat the oven to 350°F (180°C). Butter a 9-inch round cake pan.

3   In a small saucepan, combine the brown sugar, butter, cream, and cinnamon and bring to a simmer over medium heat. Simmer for 2 to 3 minutes, until you have a golden brown sauce. Pour the sauce into the prepared pan.

4   Arrange the apple slices on top of the sauce so the sauce is completely covered by the apples. Pour the cake batter on top.

5   Bake for 50 to 60 minutes, until a toothpick inserted into the center comes out clean. While the cake is still hot, run a thin knife around the edge of the cake to loosen it from the pan, then invert it onto a serving plate.

6   Let cool slightly before serving. Store leftovers in an airtight container at room temperature for up to 2 days.

DANISH DOUGH

Whenever one of my recipes is more labor-intensive than others, I make sure that the payoff is worth it. Trust me. When you take the time to make this dough and taste the results, it'll become something you don't mind taking a little longer to put together! Make it on the weekend when you have time to spare.

This dough is used to make my Sticky Maple Walnut Morning Buns (page 183), Almond Twist (page 185), and Buttery Fruit Danishes (page 186). But I encourage you to also get creative and use it for your own breakfast creations!

2½ cups (355 grams) all-purpose flour

⅓ cup (71 grams) granulated sugar

1½ teaspoons instant yeast

½ teaspoon salt

1 cup (2 sticks/225 grams) cold butter, diced

¾ cup (180 milliliters) cold whole milk

1 large egg, cold

1 In a large bowl, whisk together the flour, sugar, yeast, and salt.

2 Add the butter and, using your fingertips, rub it into the dry ingredients until the mixture resembles coarse bread crumbs.

3 In a small bowl, whisk together the milk and egg, then stir them into the flour mixture.

4 Press the dough until it just comes together. (Add a little more liquid if the dough is not holding together.)

5 Place two large pieces of plastic wrap on a work surface. Turn the dough out and use the plastic wrap to help press it into a square. Wrap well in the plastic and refrigerate for 1 hour.

6 On a floured surface, unwrap the dough and, working quickly so the butter doesn't melt, roll it out into a 12 x 18-inch rectangle. Fold it in thirds like a letter, then roll it out again to a 12 x 18-inch rectangle. Fold like a letter again, then repeat this rolling-and-folding process at least four more times, one after the other.

7 Wrap the dough well in the plastic wrap and refrigerate for at least 4 hours or up to 4 days.

8 Once chilled, use the dough as directed in the individual recipes, following the baking times and temperatures indicated.

HONEY TOFFEE SAUCE

If you are reading this and debating whether to make this sauce, the answer is YES! This honey toffee sauce is much simpler than a traditional caramel sauce, and the honey gives it an incredibly unique flavor. Make a big ol' jar; it will keep in your fridge for months on end. You will want to drizzle this on everything, believe me.

MAKES 1 CUP (283 GRAMS)

- ½ cup (1 stick/115 grams) butter
- ½ cup (115 grams) granulated sugar
- ¼ cup (71 grams) honey
- ¼ cup (60 milliliters) heavy cream
- 1 teaspoon pure vanilla extract

1 In a medium saucepan, combine the butter, sugar, and honey. Heat over medium-low heat, stirring gently, until the butter and sugar have melted, then reduce the heat to low and simmer the sauce, whisking occasionally, for 8 to 10 minutes, until it starts to darken and caramelize.

2 Remove from the heat and whisk in the cream and vanilla until combined.

3 Serve warm, or let cool and store in a jar in the fridge for up to 4 months. Reheat the cold sauce in the microwave to return it to a pourable consistency before using.

WHIPPED CREAM

I frequently get asked, "What is your recipe for whipped cream?" The answer is simple: it's cream—just a good-quality, high-fat (roughly 35%) cream. I've also included a couple of ways you can dress it up to complement the recipe it accompanies.

MAKES 2½ CUPS (590 MILLILITERS)

1½ cups (360 milliliters) cold heavy cream

Place the cream in a medium bowl and, using a handheld mixer or a whisk, whip until soft peaks form. Use immediately or cover and refrigerate until needed, up to 2 hours.

Crème Chantilly: Add 3 tablespoons granulated sugar, 2 teaspoons pure vanilla extract, and ½ teaspoon pure vanilla paste (optional) to the cream before whipping.

Coffee Cream: Dissolve 1 tablespoon instant espresso powder in 1 tablespoon hot water, then add it to the cream along with ¼ cup (57 grams) granulated sugar before whipping.

Whipped Cream

Coffee Cream

Crème Chantilly

FRUIT COMPOTES

These fruit compotes are the perfect finishing touch on my desserts. They add that extra little something that enhances flavor and adds vibrant color to a dish. I use them on a number of recipes, like my Lemon-Blueberry Ricotta Hotcakes (page 155), Baked Custard with Rhubarb Compote (page 81), and Smashed Raspberry Pavlova (page 82).

STRAWBERRY COMPOTE

MAKES 2 CUPS (282 GRAMS)

3½ cups (17½ ounces/ 497 grams) fresh strawberries, quartered

¼ cup (57 grams) granulated sugar

2 tablespoons water

2 teaspoons fresh lemon juice

1 In a medium saucepan, combine the strawberries, sugar, water, and lemon juice and bring to a very gentle simmer over medium-low heat. Cook until the berries have softened and the liquid has thickened slightly, 4 to 5 minutes. Remove from the heat and let cool completely.

2 Once cooled, use straight away or store in an airtight container in the fridge for up to 4 days.

BLUEBERRY-and LEMON COMPOTE

MAKES 2 CUPS (282 GRAMS)

3½ cups (17½ ounces/ 497 grams) fresh blueberries

¼ cup (57 grams) granulated sugar

2 tablespoons water

2 teaspoons fresh lemon juice

1 In a medium saucepan, combine the blueberries, sugar, water, and lemon juice and bring to a very gentle simmer over medium-low heat. Cook until the sugar has dissolved and the liquid has thickened slightly, 4 to 5 minutes. Remove from the heat and let cool completely.

2 Once cooled, use straight away or store in an airtight container in the fridge for up to 4 days.

Strawberry Compote

Blueberry and Lemon Compote

Rhubarb Compote

Raspberry Vanilla Compote

RHUBARB COMPOTE

MAKES 2 CUPS (282 GRAMS)

3½ cups (17½ ounces/
497 grams) chopped fresh
rhubarb (3 medium stalks)

½ cup (115 grams)
granulated sugar

2 tablespoons water

2 teaspoons fresh
lemon juice

1 In a medium saucepan, combine the rhubarb, sugar,
water, and lemon juice and bring to a very gentle
simmer over medium-low heat. Cook until the rhubarb
is tender and the liquid has thickened slightly,
5 to 6 minutes. Remove from the heat and let cool
completely.

2 Once cooled, use straight away or store in an airtight
container in the fridge for up to 4 days.

RASPBERRY VANILLA COMPOTE

MAKES 2 CUPS (282 GRAMS)

3½ cups (17½ ounces/
497 grams) fresh
raspberries

¼ cup (57 grams)
granulated sugar

2 tablespoons water

2 teaspoons fresh
lemon juice

1 vanilla bean, split in half
lengthwise

1 In a medium saucepan, combine the raspberries,
sugar, water, lemon juice, and vanilla bean and bring
to a very gentle simmer over medium-low heat.
Cook until the sugar has dissolved and the liquid has
thickened slightly, 4 to 5 minutes. Remove from the
heat and let cool completely.

2 Remove the vanilla bean. Scrape out any remaining
seeds and stir them into the compote. Use straight
away or store in an airtight container in the fridge for
up to 4 days.

CHOCOLATE-BUTTER GLAZE

Swirl, pipe, or drizzle this glaze over your desserts for another layer of rich, smooth chocolate. It's sweet but a little bitter, so it won't overpower the dessert, and silky smooth thanks to the butter. You couldn't ask for a more indulgent, balanced glaze.

MAKES ½ CUP (115 GRAMS)

- ½ cup (3 ounces/85 grams) chopped semisweet chocolate
- 4 tablespoons (½ stick/ 57 grams) butter

1 In a small microwave-safe bowl, combine the chocolate and butter. Microwave in 30-second intervals, stirring after each, to gently melt and combine. (Alternatively, melt the chocolate and butter together using a double boiler.)

2 Use immediately—pour over cakes, ice cream, or a variety of other desserts. If the glaze starts to set before use, rewarm it gently to make it pourable again.

BAKING FAQS

Salted or unsalted butter?
This is a personal choice. I use salted butter for more flavor, but that's my preference. Use whichever you prefer.

Can I use margarine instead of butter?
Yes, swapping margarine for butter works well in most recipes, but I will warn you that the quality of margarine varies and the flavor is not as good as that of real butter.

Can I use oil instead of butter?
Only when your recipe calls for melted butter; that is the only time. Butter does have a richer flavor than oil, so keep that in mind.

Can you measure liquids with measuring cups meant for dry ingredients?
I personally don't advise it. It's best to use a liquid measuring cup for liquids.

Are American cups and Australian, New Zealander, and British cups the same?
No, they are slightly different. If you are concerned about the difference, I recommend measuring ingredients by weight (grams or ounces) instead.

If I don't have baking soda, can I use baking powder instead, or vice versa?
No. Both are leavening agents, but they activate differently, so they are not interchangeable.

I don't have yeast; can I just use baking powder?
No! It is not the same thing. Sorry.

Do I need to preheat my oven?
Always. Next question.

What's the difference between a convection oven and a conventional oven?
In a conventional oven, the heat source is stationary, so hot air in the oven rises but doesn't move otherwise. This means items closer to the heating element cook faster; conventional ovens are also more prone to having "hot spots." Convection ovens have a fan that circulates hot air, distributing heat evenly inside the oven, so foods cook more consistently. The fan can be turned off, making your convection oven act like a conventional one. When do you need the fan? I personally use it when I have more than one baking sheet in the oven so the heat is distributed evenly between them.

Can milk be substituted for heavy cream in recipes?
Yes, but ONLY when the recipe doesn't call for the cream to be whipped. I do generally recommend using cream where it's called for, as it contributes extra richness and flavor.

Can I freeze eggs?
Yes! Whisk them up first and then pour them into ice-cube trays or other small containers to freeze. Then you have them ready for baking. You will need to weigh 60 grams for one large egg. Defrost in the fridge overnight before using.

Can I freeze dairy like milk, cream, and yogurt?
I really don't recommend it. When dairy defrosts, it thaws grainy and is just not the same as fresh.

Can I use hot chocolate mix as a substitute for cocoa powder?

No. Hot chocolate mix is usually made with a lower grade of cocoa powder, and it contains other ingredients like sugar, dried milk, etc. Stick to unsweetened cocoa powder for your baking.

What is the best way to melt chocolate?

This is a tricky one to answer. The *easiest* way to melt chocolate is by heating it very carefully in the microwave for a few seconds at a time. The *safest* way to melt chocolate without burning it is using a double boiler. You decide what works for you.

Can I use a chocolate candy bar for baking?

No. A candy bar is usually made with a lower grade chocolate containing little cocoa solids. It also contains added sugars. For baking, use a good-quality chocolate bar that shows the cocoa percentage on the package so you know it's the real deal.

Does humidity affect baking?

Absolutely it does. Extra moisture in the air can really affect your baked goods. If you live in a humid climate and your kitchen is warm, then work in a cooler room. Bread dough, meringue, and macarons really don't like humidity. Also, in a humid environment, your baked goods might need longer in the oven, so factor that in.

ACKNOWLEDGMENTS

Thanks a Million

In Ireland, we have a saying: "Go raibh mile maith agaibh." Simply, it means "thanks a million." And I can't say it enough to all of you who helped take my vision and bring it to life in this gorgeous book.

Ami Shukla, I can't thank you enough for your hard work and dedication and the unique culinary perspective you brought to this book and have brought to Bigger Bolder Baking in general. We are very lucky to have you as part of our team. I hope you are proud of this book, because you were a huge part of it.

To the Bigger Bolder Baking team who works behind the scenes: my mum, Alma, and Rachel, who work tirelessly answering comments and emails from bakers on a daily basis. We love you guys and appreciate your extra hard work during the time I was writing the book. Please don't ever leave me . . .

Brian, Sarah, and Zach: Thank you for keeping the ship afloat during a very busy time. As Bigger Bolder Baking grows, we are very excited for what is to come and to have you all on our team. Kevin and I are grateful for each and every one of you.

Kate Martindale, thank you for bringing your unique style, fun, and general sassiness to my book and our set. It's always a joy when we get to work together.

Carla Choy, you capture my food at its very best! All the love I put into my baking comes across in your photography. Thank you.

Olivia Crouppen, you will always be an important member of the Bigger Bolder Baking team, so I'm delighted you were able to work on book number two with me. We laughed, we sang, we danced. It's always a good time when you are around.

Maria Ribas, thank you for making the process of writing a book as easy as possible. I'm glad you are in my corner. Here's to many more books together.

Stephanie, Kerry, Jacqueline, and the rest of the Harvest team, it was a pleasure to work with you again on book number two. Thank you for giving me a blank canvas where I can be creative and do what I love to do—which is bake!

Hope Schreiber, thank you for taking my words and making them sound like poetry and not the ramblings of a crazy person. It's not an easy task, but you nailed it.

Maria, our Irish Mary Poppins, we are extremely grateful to have you in our lives and as part of our family. Thank you for all you do for us.

Sharon and Howard, your generosity toward Kevin and me over the years is unmatched. The fact that you opened your home to us to shoot my cookbook speaks volumes to the type of people you are. We are lucky to have you as neighbors but even luckier to have you as friends. Thank you.

To all my friends, family, and neighbors in Santa Monica who supported Kevin and me during this time: Whether it was to mind Georgie, eat desserts, or lend an ear, we appreciate it immensely. A special thanks to Azile, who came to take care of George from early morning until bedtime. We wouldn't have gotten through the photo shoot without you.

Georgie, after long days of writing, baking, and photography, you were always there to put a smile on my face. You are the best little boy a mammy could ask for. Love you.

Kevin, I could say so many things, but instead I'm going to keep it short. Thank you for your unwavering support. Thank you for taking a load off my hands so I could write this book. Thanks for getting up early with George and putting him down at night. Thank you for walking Waffles, even when you're exhausted. Although I went through this process myself, you were a huge part of this book's creation. I hope you love it as much as I do. Love you.

INDEX

Note: Page references in *italics* indicate photographs.

A

All-American Coffee Crumb Cake, 56, *57*
All-American Flaky Biscuits, 28, *29*
Almond flour
  Almond Twist, *184*, 185
  Ami''s Double Chocolate Almond Cookies, 60, *61*
Almond(s)
  Crisp Topping, 226, *227*
  10-Minute Summer Berry Tiramisu, 68, *69*
  Twist, *184*, 185
Amaretto Cherry Clafoutis, 199, *200*
Ami's Double Chocolate Almond Cookies, 60, *61*
Apple(s)
  Buttery Fruit Danishes, 186, *187*
  Cake, Mum's Irish, 52, *53*
  Caramel, Crostata, 207, *208*
  Cinnamon Scones, 4, *5*
  Cinnamon Upside-Down Cake, Fall, *213*, 216
  Crisp, Traditional, 195, *196*
  Moist Fruit-and-Nut Cake, 50, *51*
  Oatmeal Muffins, *26*, 27
Apricots and apricot jam
  Buttery Fruit Danishes, 186, *187*
  Whole Wheat and Fruit Breakfast Bread, *32*, 33
Aussie Pikelets, Georgie's, *156*, 157

B

Babka, Braided Chocolate, 166–69, *167*
Bakery-Style Lemon Blueberry Muffins, 14, *15*
Baking conversion chart, xiv–xv
Baking FAQs, 238–39
Baking pan sizes, xvii
Baking tips, xiii

Banana(s)

Bread Muffins, 20, *21*
Cake, Butterscotch, 62, *63*
Caramelized, with Rosemary, *203*, 205
Cinnamon Waffles, *140*, 141
Oat Pancakes, Kevin's No-Fuss, *146*, 147
Pudding, Old-Fashioned, 70, *71*
and Rum Upside-Down Cake, *213*, 215
Belgian Waffles, Overnight, *138*, 139
Berry(ies). *See also specific berries*
  and Cream Cheese Crostata, *208*, 209
  Summer, Tiramisu, 10-Minute, 68, *69*
  Triple, Crisp, 195, *196*
Biscuits
  All-American Flaky, 28, *29*
  Buttermilk Drop, 30, *31*
Blackberry
  Clafoutis, *200*, 201
  Fool, Summer, *193*, 194
Blueberry(ies)
  Compote, Chèvre and Honey Tart with, 112, *113*
  and Lemon Compote, 234, *235*
  Lemon Muffins, Bakery-Style, 14, *15*
  -Lemon Ricotta Hotcakes, *154*, 155
  Peach Crisp, *196*, 197
  Pie, Classic, 88, *89*
  and Rose Water Fool, 192, *193*
  10-Minute Summer Berry Tiramisu, 68, *69*
Boozy Chocolate and Prune Cake, *114*, 115
Boozy Raspberry Fool, *193*, 194
Braided Chocolate Babka, 166–69, *167*
Bread(s)
  -and-Butter Pudding, Traditional Irish, *86*, 87
  Braided Chocolate Babka, 166–69, *167*
  Caramel Pecan Monkey, 170–73, *171*
  Carrot and Marmalade, 34, *35*
  Cinnamon Raisin, No-Knead, 162, *163*
  Make-Ahead French Toast Casserole, 150, *151*
  My Go-To Brioche Loaf, *174*, 175
  Sweet Pumpernickel, *164*, 165
  Whole Wheat and Fruit Breakfast, *32*, 33

Brioche Buns, Breakfast, *176, 177*
Brioche Loaf, My Go-To, *174,* 175
Buns
  Breakfast Brioche, *176,* 177
  Sticky Maple Walnut Morning, *182,* 183
Butter-Chocolate Glaze, 237
Buttermilk Drop Biscuits, 30, *31*
Butterscotch Banana Cake, 62, *63*
Buttery Fruit Danishes, 186, *187*

C

Cakes
  Apple, Mum's Irish, 52, *53*
  Apple Cinnamon Upside-Down, Fall, *213,* 216
  Banana and Rum Upside-Down, *213,* 215
  Butterscotch Banana, 62, *63*
  Cappuccino Swiss Roll, 46, *47*
  Chocolate, Squidgy, 84, *85*
  Chocolate and Prune, Boozy, *114,* 115
  Citrus Olive Oil Pound, *54,* 55
  Coffee Crumb, All-American, 56, *57*
  Death by Chocolate, 126–28, *127*
  Dulce de Leche Lava, 118, *119*
  French Yogurt Pot, *58, 59*
  Fruit-and-Nut, Moist, 50, *51*
  Maple and Pear Upside-Down, 212, *213*
  Orange and Honey Toffee Upside-Down, *213,* 214
  Polenta, with Mascarpone and Strawberry Compote, *64, 65*
  Strawberry Dump, *96, 97*
  Tiramisu Crepe, Elegant, *110,* 111
Cappuccino Swiss Roll, 46, *47*
Caramel Apple Crostata, 207, *208*
Caramel Pecan Monkey Bread, 170–73, *171*
Carrot
  Cake Pancakes with Cream Cheese Frosting, *144,* 145
  and Marmalade Bread, 34, *35*
Cheese. *See also* Cream Cheese; Mascarpone
  Chèvre and Honey Tart with Blueberry Compote, 112, *113*
  Lemon-Blueberry Ricotta Hotcakes, *154,* 155

Cheesecakes
  Chocolate Lover's, with Strawberry Compote, *78,* 79
  White Chocolate and Passion Fruit, 104, *105*
Cherry(ies)
  Amaretto Clafoutis, 199, *200*
  Macerated, Yogurt Coeur à la Crème with, 106, *107*
Chèvre and Honey Tart with Blueberry Compote, 112, *113*
Chocolate. *See also* White Chocolate
  Babka, Braided, 166–69, *167*
  -Butter Glaze, 237
  Cake, Death by, 126–28, *127*
  Cake, Squidgy, *84,* 85
  Cappuccino Swiss Roll, 46, *47*
  Chip Pumpkin Muffins, *24,* 25
  Chocolate Lover's Cheesecake with Strawberry Compote, *78,* 79
  Dark, and Pear Crisp, Unbelievable, 132, *133*
  Decadent Cocoa Panna Cotta, 108, *109*
  Double, Almond Cookies, Ami's, 60, *61*
  Éclairs, Classic, 44, *45*
  Hazelnut, and Mascarpone Crostata, 206, *208*
  -Oat Tea Cookies, Homemade, 42, *43*
  and Prune Cake, Boozy, *114,* 115
  Swirl Meringues, 48, 49
  Triple, Muffins, 16, *17*
Cinnamon
  All-American Coffee Crumb Cake, 56, *57*
  Apple Scones, 4, *5*
  Apple Upside-Down Cake, Fall, *213,* 216
  Banana Waffles, *140,* 141
  Caramel Pecan Monkey Bread, 170–73, *171*
  Raisin Bread, No-Knead, 162, *163*
  Rolls, No-Yeast, 136, *137*
  Semifreddo with Honey Toffee Swirl, *94,* 95
  Sour Cream Coffee Cake Muffins, 18, *19*
  Sticky Maple Walnut Morning Buns, *182,* 183
Citrus Olive Oil Pound Cake, *54,* 55
Clafouti Batter, 198, *198*
Clafoutis
  Blackberry, *200,* 201
  Cherry Amaretto, 199, *200*

Fresh Fig and Cardamom, *200,* 201

Peach and Raspberry, 199, *200*

Cobbler, Pecan Pie, 98, *99*

Coconut

Crème Brûlée, 122, *123*

Rum Sauce, Roasted Pineapple with, 202, *203*

Coeur à la Crème, Yogurt, with Macerated
Cherries, 106, *107*

Coffee

Cappuccino Swiss Roll, *46,* 47

Elegant Tiramisu Crepe Cake, *110,* 111

Pudding, Silky, *120,* 121

Whipped Cream, 232, *233*

Coffee Crumb Cake, All-American, 56, *57*

Compotes

Blueberry and Lemon, 234, *235*

Raspberry Vanilla, *235,* 236

Rhubarb, *235,* 236

Strawberry, 234, *235*

Cookie(s)

Buttery Shortbread, Mia's, *38, 39*

Chocolate-Oat Tea, Homemade, 42, *43*

Double Chocolate Almond, Ami's, 60, *61*

Linzer, Authentic, *40,* 41

White Chocolate Pecan Skillet, The
Ultimate, 90, *91*

Cornmeal. *See* Johnnycakes

Cream Cheese

and Berry Crostata, *208,* 209

Breakfast Brioche Buns, *176,* 177

Carrot and Marmalade Bread, 34, *35*

Chèvre and Honey Tart with Blueberry Compote,
112, *113*

Chocolate Lover's Cheesecake with Strawberry
Compote, *78,* 79

Frosting, Carrot Cake Pancakes with, *144,* 145

No-Yeast Cinnamon Rolls, 136, *137*

and Raspberry Crepe Casserole, 158, *159*

Strawberry Scones, *12,* 13

White Chocolate and Passion Fruit Cheesecake,
104, *105*

Crème Brûlée, Coconut, 122, *123*

Crème Chantilly, 232, *233*

Crème Fraîche, Plum, and Vanilla Bean Crostata,
208, 210

Crepe Cake, Elegant Tiramisu, *110,* 111

Crepe Casserole, Raspberry and Cream
Cheese, 158, *159*

Crepes, 225

Crisps

Apple, Traditional, 195, *196*

Blueberry Peach, *196,* 197

Crisp Topping for, 226, *227*

Pear and Dark Chocolate, Unbelievable,
132, *133*

Rhubarb and Strawberry, *196,* 197

Triple Berry, 195, *196*

Croissants, Easy-Peasy Homemade,
178–81, *179*

Crostatas

Berry and Cream Cheese, *208,* 209

Caramel Apple, 207, *208*

Chocolate, Hazelnut, and Mascarpone, 206, *208*

Plum, Crème Fraîche, and Vanilla Bean, *208,* 210

Custards

Baked, with Rhubarb Compote, *80,* 81

Coconut Crème Brûlée, 122, *123*

D

Danish Dough, 228, *229*

Danishes, Buttery Fruit, 186, *187*

Dates

Moist Fruit-and-Nut Cake, 50, *51*

Death by Chocolate Cake, 126–28, *127*

Drop Biscuits, Buttermilk, 30, *31*

Dulce de Leche Lava Cake, 118, *119*

Dutch Baby Pancake, 5-Ingredient, 142, *143*

E

Éclairs, Classic Chocolate, 44, *45*

F

Fig(s)

Fresh, and Cardamom Clafoutis, *200,* 201

Honey Glazed, with Whipped Mascarpone,
203, 205

5-Ingredient Dutch Baby Pancake, 142, *143*

Fools
    Blueberry and Rose Water, 192, *193*
    Boozy Raspberry, *193*, 194
    Strawberry and Mint, 192, *193*
    Summer Blackberry, *193*, 194
French Toast Casserole, Make-Ahead, 150, *151*
French Yogurt Pot Cake, *58,* 59
Fruit. *See also* Berry(ies); *specific fruits*
    -and-Nut Cake, Moist, 50, *51*
    Compotes, 234, *235*
    Crisps, 195–97, *196*
    Danishes, Buttery, 186, *187*
    Fools, 192–94, *193*
    Pan-Roasted, 202–5, *203*
    Summer, and Rose, Layered Pavlova with, *130,* 131
    and Whole Wheat Bread, *32,* 33
    and Whole Wheat Breakfast Bread, *32,* 33

G

Georgie's Aussie Pikelets, *156,* 157
Glaze, Chocolate-Butter, 237
Granita, Sparkling Rosé and Raspberry, 116, *117*

H

Hawaiian Sweet Rolls, 188, *189*
Hazelnut, Chocolate, and Mascarpone Crostata, 206, *208*
Honey
    Glazed Figs with Whipped Mascarpone, *203,* 205
    Toffee and Orange Upside-Down Cake, *213,* 214
    Toffee Sauce, 230, *231*
    Toffee Swirl, Cinnamon Semifreddo with, *94,* 95

I

Irish Apple Cake, Mum's, 52, *53*
Irish Bread-and-Butter Pudding, Traditional, *86,* 87
Irish Scones, Traditional, *2,* 3

J

Jam. *See* Raspberry jam
Johnnycakes, Old-Fashioned, *152,* 153

L

Ladyfingers
    10-Minute Summer Berry Tiramisu, 68, *69*
Lava Cake, Dulce de Leche, 118, *119*
Lemon
    Blueberry Muffins, Bakery-Style, 14, *15*
    -Blueberry Ricotta Hotcakes, *154,* 155
    Citrus Olive Oil Pound Cake, *54,* 55
    Whole, Tart, 76, *77*
Lime
    Coconut Crème Brûlée, 122, *123*
    Custard Pie, Cool and Creamy, 92, *93*
Linzer Cookies, Authentic, *40,* 41

M

Make-Ahead French Toast Casserole, 150, *151*
Maple
    and Pear Upside-Down Cake, 212, *213*
    Pecan Scones, 8, *9*
    Walnut Morning Buns, Sticky, *182,* 183
Marmalade
    and Carrot Bread, 34, *35*
    Pudding, Steamed, *72,* 73
Mascarpone
    Chocolate, and Hazelnut Crostata, 206, *208*
    Elegant Tiramisu Crepe Cake, *110,* 111
    and Strawberry Compote, Polenta Cake with, *64,* 65
    10-Minute Summer Berry Tiramisu, 68, *69*
    Whipped, Honey Glazed Figs with, *203,* 205
    Yogurt Coeur à la Crème with Macerated Cherries, 106, *107*
Meringue(s)
    Basic, 220, *221*
    Chocolate Swirl, 48, *49*
    Layered Pavlova with Summer Fruit and Rose, *130,* 131
    Smashed Raspberry Pavlova, 82, *83*

Mia's Buttery Shortbread Cookies, *38,* 39
Mint and Strawberry Fool, 192, *193*
Monkey Bread, Caramel Pecan, 170–73, *171*
Muffins
  Apple Oatmeal, *26,* 27
  Banana Bread, 20, *21*
  Lemon Blueberry, Bakery-Style, 14, *15*
  Pumpkin Chocolate Chip, *24,* 25
  Sour Cream Coffee Cake, 18, *19*
  Triple Chocolate, *16,* 17
Mum's Irish Apple Cake, 52, *53*

N

No-Knead Cinnamon Raisin Bread, 162, *163*
No-Yeast Cinnamon Rolls, 136, *137*
Nuts. *See specific nuts*

O

Oat(s)
  Apple Oatmeal Muffins, *26,* 27
  Banana Pancakes, Kevin's No-Fuss, *146,* 147
  -Chocolate Tea Cookies, Homemade, 42, *43*
  Crisp Topping, 226, *227*
  Whole Wheat and Fruit Breakfast Bread, *32,* 33
Olive Oil Citrus Pound Cake, *54,* 55
Orange(s)
  Citrus Olive Oil Pound Cake, *54,* 55
  and Honey Toffee Upside-Down Cake, *213,* 214
  Steamed Marmalade Pudding, *72,* 73
Oven temperature chart, xvi
Overnight Belgian Waffles, *138,* 139

P

Pancake, 5-Ingredient Dutch Baby, 142, *143*
Pancakes
  Banana Oat, Kevin's No-Fuss, *146,* 147
  Carrot Cake, with Cream Cheese Frosting, *144,* 145
  Georgie's Aussie Pikelets, *156,* 157
  Lemon-Blueberry Ricotta Hotcakes, *154,* 155

Panna Cotta, Decadent Cocoa, 108, *109*
Pan-Roasted Fruit, 202–5, *203*
Passion Fruit and White Chocolate Cheesecake, 104, *105*
Pastries
  Almond Twist, *184,* 185
  Buttery Fruit Danishes, 186, *187*
  Classic Chocolate Éclairs, 44, *45*
Pastry dough
  Danish Dough, 228, *229*
  Pie Crust, 222, *223*
Pavlova
  Layered, with Summer Fruit and Rose, *130,* 131
  Smashed Raspberry, 82, *83*
Peach(es)
  Blueberry Crisp, *196,* 197
  Layered Pavlova with Summer Fruit and Rose, *130,* 131
  and Raspberry Clafoutis, 199, *200*
  Shortcakes, Flaky, *74,* 75
Pear
  and Dark Chocolate Crisp, Unbelievable, 132, *133*
  and Maple Upside-Down Cake, 212, *213*
Pecan(s)
  All-American Coffee Crumb Cake, 56, *57*
  Caramel Monkey Bread, 170–73, *171*
  Maple Scones, 8, *9*
  Pie Cobbler, 98, *99*
  White Chocolate Skillet Cookie, The Ultimate, 90, *91*
Pie Crust, 222, *223*
Pies
  Blueberry, Classic, 88, *89*
  Lime Custard, Cool and Creamy, 92, *93*
Pikelets, Georgie's Aussie, *156,* 157
Pineapple, Roasted, with Rum Coconut Sauce, 202, *203*
Plum, Crème Fraîche, and Vanilla Bean Crostata, *208,* 210
Polenta Cake with Mascarpone and Strawberry Compote, *64,* 65
Popovers, Classic, *148,* 149
Pound Cake, Citrus Olive Oil, *54,* 55

Prune and Chocolate Cake, Boozy, *114,* 115
Pudding
  Banana, Old-Fashioned, 70, *71*
  Bread-and-Butter, Traditional Irish, *86,* 87
  Coffee, Silky, *120,* 121
  Marmalade, Steamed, *72, 73*
Pumpernickel Bread, Sweet, *164,* 165
Pumpkin Chocolate Chip Muffins, *24,* 25

R

Raisin(s)
  Buttery Fruit Danishes, 186, *187*
  Cinnamon Bread, No-Knead, 162, *163*
  Rum Semifreddo, Grown-Ups', 124, *125*
  Traditional Irish Bread-and-Butter
    Pudding, *86,* 87
  Traditional Irish Scones, *2,* 3
  Whole Wheat and Fruit Breakfast Bread, *32, 33*
Raspberry(ies)
  and Cream Cheese Crepe Casserole, 158, *159*
  Fool, Boozy, *193,* 194
  and Peach Clafoutis, 199, *200*
  Smashed, Pavlova, 82, *83*
  and Sparkling Rosé Granita, 116, *117*
  10-Minute Summer Berry Tiramisu, 68, *69*
  Vanilla Compote, *235, 236*
  and Yogurt Scones, *6,* 7
Raspberry jam
  Authentic Linzer Cookies, *40, 41*
  Breakfast Brioche Buns, *176,* 177
  French Yogurt Pot Cake, *58, 59*
  Gooey Jam Tart, 100, *101*
Rhubarb
  Compote, *235,* 236
  Compote, Baked Custard with, *80,* 81
  and Strawberry Crisp, *196,* 197
Ricotta Lemon-Blueberry Hotcakes, *154,* 155
Rolls
  Cinnamon, No-Yeast, 136, *137*
  Hawaiian Sweet, 188, *189*

Rosemary, Caramelized Bananas with, *203,* 205
Rose Water
  and Blueberry Fool, 192, *193*
  Layered Pavlova with Summer Fruit and Rose,
    130, 131
  Roasted Strawberries with Rose and Cardamom
    Yogurt, *203,* 204
Rum
  and Banana Upside-Down Cake, *213,* 215
  Boozy Chocolate and Prune Cake, *114,* 115
  Coconut Sauce, Roasted Pineapple with,
    202, *203*
  Elegant Tiramisu Crepe Cake, *110,* 111
  Raisin Semifreddo, Grown-Ups', 124, *125*

S

Sauce, Honey Toffee, 230, *231*
Scones
  Cinnamon Apple, 4, *5*
  Maple Pecan, 8, *9*
  Raspberry and Yogurt, *6,* 7
  Strawberry Cream Cheese, *12,* 13
  Three-Seed Whole Wheat, 10, *11*
  Traditional Irish, *2,* 3
Seeds, Three-, Whole Wheat Scones, 10, *11*
Semifreddo
  Cinnamon, with Honey Toffee Swirl, *94,* 95
  Rum Raisin, Grown-Ups', 124, *125*
Shortbread Cookies, Mia's Buttery, *38,* 39
Shortcakes, Flaky Peach, *74, 75*
Sour Cream Coffee Cake Muffins, 18, *19*
Sparkling Rosé and Raspberry Granita, 116, *117*
Steamed Marmalade Pudding, *72, 73*
Sticky Maple Walnut Morning Buns, *182,* 183
Strawberry(ies)
  Compote, 234, *235*
  Compote, Chocolate Lover's Cheesecake
    with, *78, 79*
  Compote and Mascarpone, Polenta Cake
    with, *64,* 65
  Cream Cheese Scones, *12,* 13

Dump Cake, *96,* 97

5-Ingredient Dutch Baby Pancake, 142, *143*

Layered Pavlova with Summer Fruit and Rose, *130,* 131

and Mint Fool, 192, *193*

and Rhubarb Crisp, *196,* 197

Roasted, with Rose and Cardamom Yogurt, *203,* 204

10-Minute Summer Berry Tiramisu, 68, *69*

Swiss Roll, Cappuccino, *46,* **47**

T

Tarts. *See also* **Crostatas**

Chèvre and Honey, with Blueberry Compote, 112, *113*

Gooey Jam, 100, *101*

Whole Lemon, 76, *77*

10-Minute Summer Berry Tiramisu, 68, *69*

Three-Seed Whole Wheat Scones, 10, *11*

Tiramisu, 10-Minute Summer Berry, 68, *69*

Tiramisu Crepe Cake, Elegant, *110,* **111**

Toffee

Honey, and Orange Upside-Down Cake, *213,* 214

Honey Sauce, 230, *231*

Honey Swirl, Cinnamon Semifreddo with, *94,* 95

Triple Chocolate Muffins, *16,* **17**

U

Upside-Down Cake Batter, 211, *211*

Upside-Down Cakes

Banana and Rum, *213,* 215

Fall Apple Cinnamon, *213,* 216

Maple and Pear, 212, *213*

Orange and Honey Toffee, *213,* 214

W

Waffles

Banana Cinnamon, *140,* 141

Belgian, Overnight, *138,* 139

Walnut(s)

Cinnamon Semifreddo with Honey Toffee Swirl, *94,* 95

Maple Morning Buns, Sticky, *182,* 183

Moist Fruit-and-Nut Cake, 50, *51*

Sour Cream Coffee Cake Muffins, 18, *19*

Whipped Cream, 232, *233*

Coffee Cream, 232, *233*

Crème Chantilly, 232, *233*

Fruit Fools, 192–94, *193*

White Chocolate

and Passion Fruit Cheesecake, 104, *105*

Pecan Skillet Cookie, The Ultimate, 90, *91*

Whole Wheat and Fruit Breakfast Bread, *32,* **33**

Whole Wheat Scones, Three-Seed, 10, *11*

Y

Yogurt

Coeur à la Crème with Macerated Cherries, 106, *107*

French Yogurt Pot Cake, *58,* 59

and Raspberry Scones, 6, *7*

Rose and Cardamom, Roasted Strawberries with, *203,* 204